Looking east, looking west

Looking east, looking west

Organic and quality food marketing in Asia and Europe

edited by:
Rainer Haas
Maurizio Canavari
Bill Slee
Chen Tong
Bundit Anurugsa

Wageningen Academic
P u b l i s h e r s

This publication is a result of the BEAN-QUORUM project that was supported by the European Union in 2004-2008

ASIA LINK

EuropeAid
CO-OPERATION OFFICE

ISBN: 978-90-8686-095-1
e-ISBN: 978-90-8686-703-5
DOI: 10.3921/978-90-8686-703-5

First published, 2010

© Wageningen Academic Publishers
The Netherlands, 2010

Table of contents

Preface

Consolidating cooperation among Asia and Europe on quality, organic and unique food marketing

By Friedrich Hamburger, ambassador and head of Delegation of the European Commission to Thailand in 2008

Many know the EU as the largest donor of development assistance worldwide. The Union also promotes trade to drive development by ensuring its markets are open to exports from emerging economies and less developed countries. Over 60% of exports from emerging and least developed countries head for the EU while Europe imports more agricultural products from developing countries than the US, Japan and Canada combined.

Increased consumer awareness of food safety issues and environmental concerns have contributed to the growth in organic farming as well as markets for organic produce over recent years. In the European Union, organic farming has in fact developed into one of the most dynamic agricultural sectors.

In 2005, in the European Union of 25 Member States, around 6 million hectares were either farmed organically or were being converted to organic production. This marks an increase of more than 2% compared with 2004. Over the same period, the number of organic operators grew by more than 6%. The organic farm sector grew by about 25% a year between 1993 and 1998 and, since 1998, is estimated to have grown by around 30% a year. This rise of organic farming in Europe has been triggered by a considerable growth in the market for organic products in the EU in recent years. Currently almost half of all organic products worldwide are sold in Europe, and there is no sign yet of an end to this increase. Even if at present only 3.2% of the European agricultural land area is managed organically, and the market share of organic produce is between 1 and 2%, organic agriculture has secured a place in the economy and society in Europe.

The EU is also leading the way on developing regulations concerning organic production and labelling, as the environment, health and food safety is an issue close to the heart of many European citizens. They feel strongly about the kind of environment they want to live in and about the environment they want their children and grandchildren to inherit. This concern is extended to the type of products they wish to use or consume. Therefore, a product that is able to prove that it is better for the environment and health by a trustworthy label, can help make a difference in the eyes of European consumers.

In June 2007, as some of you will know, the European Union reached a political agreement on a new regulation on organic production and labelling. The new regulation will be effective from 2009. It sets out the new objectives, principles and basic rules for organic production in Europe. It also includes new permanent import and control regimes. Organic food is a successful and growing market and I hope that this new set of rules will provide the framework to allow this growth to continue; through a combination of market demand and the entrepreneurship of European farmers. Important from a consumers' point of view, the new regulation will make it easier for consumers to recognize organic products and it will provide assurance of precisely what is being bought.

In addition to regulatory aspects the European Union has also been active in many countries, providing funding for food quality, safety standards and related activities. By delivering short-term expertise in a flexible manner through the Asian Trust Fund (ATF), the European Union has contributed to raising the institutional capacity to deal with complex trade negotiations and issues as well as opportunities, such as that presented by the quality of agriculture products.

One of the projects supported under the ATF provided assistance to Thailand to establish a control system for organic products so that the country could eventually apply for inclusion in the European Union's 'Third country list' of exporting countries. European experts worked with national authorities to develop the 7-point national action plan for Thailand's organic sector. The plan covers not only the production chain but also the aspects of regulations, certification, research, training and marketing.

Other food-related activities the EU has been involved in with Thailand include: the EC-ASEAN intellectual property rights cooperation programme (ECAP II). One of the key activities undertaken by this initiative has been the promotion of Geographical indications for food products. The recently completed EU-Thailand Small Projects Facility (SPF) funded actions on safe shrimp production and the setting up of a regional rapid alert system for food products amongst others. While at Thammasat University, European experts have joined efforts to enhance competitiveness of Thai organic jasmine rice and tapioca by transferring technical expertise to local farmers.

Finally, I would like to inform you about the new SWITCH-Asia Programme. The overall objective of this new programme is to promote economic prosperity and poverty reduction in Asian countries through sustainable growth with reduced environmental impact by industries and consumers, in line with international environmental agreements and treaties. The *purpose* is to promote sustainable production (i.e. development of less polluting and more resource efficient products, processes and services) and sustainable consumption patterns and behaviour in the Asia region, through an improved reciprocal understanding and strengthened cooperation between Europe and Asia, notably by mobilizing the private sector, i.e. SME's, retailers and consumer groups/organisations. This programme, alongside the 7th Research framework programme and the new version of the Small projects facility, in 2007 renamed Thailand-EC Cooperation Facility (TCF), represents the most interesting means for possible future funding from the European Union.

I understand, nevertheless, that substantial momentum has been achieved by the BEAN-QUORUM network. And I certainly expect that there will be an abundance of opportunities to continue the good work and extending the network yet further. I am confident therefore to say that the EU has considerable expertise to share in this field. Yet, interest and expertise in quality of agriculture products and organic farming in Asia has also grown substantially over the past decade.

Khob Khun Krap
(Thai: thank you)

Friedrich Hamburger

Acknowledgements

This text is the reprint of the speech held by Dr. Friedrich Hamburger at the Thammasat University, Rangsit Campus, during the BEAN-QUORUM Asia-Link dissemination symposium held on the 8th of February 2008.

Organic and quality food marketing in Asia and Europe: a double sided perspective on marketing of quality food products

R. Haas, M. Canavari, B. Slee, T. Chen and B. Anurugsa

1. Subject matter

This book represents a unique collection of European and Asian perspectives on the production, trade and consumption of high quality food. It has been motivated by the common interest of a team of researchers and teachers in the field of quality food marketing. Overall improvement of the quality of agri-food products is a strategic task for agriculture and rural economic development and has become a priority of companies and general policy in many countries (Steenkamp, 1990).

Food quality is a major topic in the scientific debate over agri-food products, although in discussions on international trade, as well as when talking of agri-food marketing from the macroeconomic point of view, quality is usually a neglected aspect, since the analysis is often focused on commodities, i.e. undifferentiated goods. In fact, in international trade the problem of quality is usually resolved by standardization, while in macroeconomic analysis of agricultural markets the assumption is often that produce is typically homogeneous in nature.

However, the importance of marketing and trade in quality food products both for developed and developing countries is growing. The rapidly growing demand for organic and quality food in Europe imposes new challenges on competing food value chains. Europe, as the biggest worldwide food importer, attracts many developing and developed countries in Asia. Prospering Chinese and Thai food markets offer new opportunities for European operators. Affluent and informed consumers on both continents search for trustworthy high quality food products. Farmers, operators and retailers from distant cultures are coping with different standards, facing the ever-increasing necessity for mutual understanding.

Since an important element of a marketing strategy based on quality is consumers' perceptions of quality, perceived quality judgments are crucial and it is necessary to understand the role of personal (e.g. psychological, cultural) and situational variables, together with the perception process itself and the product/service attributes, in the formation of quality judgments (Oude Ophuis and Van Trijp, 1995). Therefore, a recent strand of the economic literature recognizes that, for instance, local/traditional food specialties, as well as organic food products, may represent a starting point in enhancing the value of food production also in the less developed countries, allowing them to gain access to interesting markets or to provide a higher value products on domestic markets, and thus promoting economic and social development.

On the demand side, consumer preferences and increased purchasing power enable the emergence of meaningful and actionable market segments, thus increasing the need for products that are differentiated on the basis of their unique quality characteristics, which are related to sensory, cultural, functional, ethical, and other attributes that are desirable for some consumers.

The concept of quality is evocative of what is 'good' or 'better' or 'different'. However, it is a very complex concept when analyzed in depth. The variety of quality definitions that are proposed in several fields show that this concept is actually very fragmented and multifaceted. A short and effective definition of quality is the ability of an entity (e.g. product or service) to satisfy implicit or explicit needs, or the 'degree to which a set of inherent characteristics fulfils requirements'

(ISO, 2005) or in simpler words 'we keep our promises'. For food quality may be defined as 'everything a consumer would find desirable in a food product' (Grunert, 2005). These definitions imply that provision of quality for a supplier involves meeting customer requirements and desires or, in marketing parlance, meeting the needs, wants and expectations of customers. Furthermore, it therefore makes the concept of quality, being dependent on customer preference, highly subjective. This explains why there is an intimate link between marketing and quality.

The specificity of the food industry has highlighted the need to develop a conceptual approach that could be more consistent with the peculiarities of both the organization of the food sector and the behavior of food product consumers. The publication of numerous studies on consumer preferences and behavior considering quality and value-adding attributes of food is a demonstration that this link is already a cornerstone of recent research in agricultural and food economics. Efforts have been made in this direction, for instance by Brunsø *et al.* (2002) with the Total quality food model.

The volume contains contributions aimed at addressing many facets related to quality in the agrifood industry and in different international contexts. This publication is the output of BEAN-QUORUM a European funded Asia-Link project, which is briefly described in the following section.

2. The BEAN-QUORUM project

The original research that led to the papers published here was developed within the framework of the BEAN-QUORUM (Building a European-Asian Network for Quality, Organic, and Unique food Marketing) network, which is the final result of a project funded within the framework of the five-year Asia-link programme. The Asia link was set up by the European Commission in 2002 so as to promote regional and multilateral networking between higher education institutions in Europe and developing countries in Asia. The programme had a budget of 42.8 million euro and was aimed at providing support to European and Asian higher education institutions in the areas of human resource development, curriculum development and institutional and systems development.

The BEAN-QUORUM project (www.bean-qourum.net) started on January 1, 2005 and ended on May 31, 2008 following a 5-months extension granted by the EC Delegation in Thailand. The project was aimed at creating a network of Asian and European higher education institutions that are interested in the marketing issues regarding quality food and organic food products.

The BEAN-QUORUM partnership was formed by the following academic partners:
- Department of Agricultural Economics and Engineering, Alma Mater Studiorum University of Bologna, Italy
- The Macaulay Land Use Research Institute, United Kingdom (which substituted the University of Gloucestershire after the first year)
- Xinjiang Agricultural University, P.R. of China
- Faculty of Science and Technology, Thammasat University, Thailand
- University of Natural Resources and Applied Life Sciences, Vienna, Austria

After the end of the project access to partnership was open to those Universities or research centers that showed an interest in the topic of the thematic network. To date the new academic partners are the following:
- Shanghai Jiao Tong University (P.R. of China)

- Nanjing Agricultural University (P.R. of China)
- Technical University of Madrid (Spain)
- Himalayan College of Agricultural Sciences and Technology (Nepal)
- Centre for Agrarian Systems Research and Development - CASRAD (Vietnam)

Several non-academic partners also joined the network as associates, thus demonstrating the lively interest that this initiative raised in the food-related industry and governmental operators.

During the 41-month duration of the project, 4 main events characterized the project life-cycle. In the first year the Xinjiang kick-off meeting (18-24 July 2005) was very important to fine-tune the plans for all the activities enabling us to reach our logical framework target indicators. The first meeting of the partnership participants was aimed at creating a basis for further collaboration during the project development period and starting up the project activities. The several points of view of the partners regarding procedures, market conditions in the quality and organic food markets and mainstream marketing approaches and their application in the local markets were the main topics of the discussion.

The meeting in Bangkok (Thailand), held in February-March 2006, was the first crucial step aimed at reaching important project teaching aims. The Short Intensive Overseas Programme on Organic Agriculture held in Bangkok in 2006 involved 25 researchers who benefited from inputs from a range of European, Chinese and local experts, covering all aspects of organic and quality food certification and marketing and rural development. This short course was the basis for establishing arrangements for mutual credit recognition by the university partners to help encourage future exchanges.

The BEAN-QUORUM partners provided teaching and organization resources and Thammasat University provided the course facilities, organization and also the attendees to the course. This 2-week training course, aimed at transferring knowledge about marketing of organic agri-food products and the interaction between international teachers and local graduate students, scholars, and researchers was highly productive.

The meeting in Bologna (3-10 March 2007) was the research-focused event in the project action plan. During that week many activities were undertaken, but the most significant was the 105th Seminar on 'International marketing and international trading of quality food products' that involved more than 100 scholars from 20 countries.

The last meeting was held at Thammasat University from the 5th of February to the 12th of February 2008 An open conference with the theme 'Consolidating co-operation between Europe and Asia on quality, organic and unique food marketing' was held at Thammasat University's Rangsit Campus on Friday 8 February to mark the conclusion of the EU-funded phase of the BEAN-QUORUM project. This conference was aimed at strengthening international cooperation ties between the partners and to put them in touch with public bodies representatives and market operators, who provided a wide set of information and helped in communicating the project results and the anticipated further developments to the public. The EU ambassador to Thailand, His Excellency Mr. Hamburger, gave a presentation pointing out the relevance of the project to the Commission's recent thinking on development in the food sector.

Cross-cutting activities between these events were undertaken:
- Multilateral networking: a memorandum of understanding (multilateral network agreement) has been agreed upon and signed by the relevant legal representatives of the five academic

partners of the BEAN-QUORUM project. Additional partners, both academic and associate, have also officially joined the network, while others showed their interest. The BEAN-QUORUM partnership alongside the BEAN-QUORUM network website, has started to manage a thematic social network on the Internet, aimed at enabling also individual scholars or practitioners to contribute to the life of the network.

- Development of industrial relationships: contacts with companies and associations who have shown interest in the activities of the BEAN-QUORUM network happened throughout the whole project's life-cycle. In order to foster the networking and industrial relationships, several contacts with industrial operators have been actively promoted in order to raise interest in the network. Meetings and round tables with industry operators and practitioners have been organised during the Xinjiang, Bangkok, and Bologna meetings. Participation in a networking meeting in Vienna also took place.

- Implementation of training activities: besides the SIOP course in Bangkok (by Thammassat University) integration of the BEAN-QUORUM activities in teaching courses at the partner Universities and the Bioagricert Training for International Inspectors in Casalecchio di Reno (Bologna, Italy) also took place. A curriculum for a masters program in Organic Agriculture Management has been developed and approved by Thammasat University. The E-learning portal developed under the guidance of professor Rainer Haas from BOKU, uses an open source programme to provide an archive of all the learning material accumulated during the project. This software offers a wide range of material to the expanding BEAN-QUORUM network. Really valuable information can flow around the network without the movement of people. E-learning methods are widely used in European universities and sharing these approaches within the network was a highly valuable experience.

- Staff exchanges between universities and study/research collaboration: eight young researchers worked for three to four months in different partner institutions. Two researchers from Austria and China went to Thailand and researched the possible income benefits of organic farming in Thailand and the marketing of organic products in Thailand. Two researchers from China and Thailand visited Bologna and worked on market research projects. Two researchers from Thailand and China visited the Macaulay institute in Aberdeen. Another researcher from China visited Vienna and investigated the marketing of organic food in Austria. Furthermore, a researcher from Italy went to China and performed research on the interest of Chinese consumers for organic products and on the interest of Chinese distribution practitioners in Italian food specialties. In addition, the EAAE seminar mentioned above has given rise to the publication of a proceedings CD-ROM which contained all the contributions presented by network participants (Chai *et al.*, 2007; Marchesini *et al.*, 2007; Morawetz *et al.*, 2007; Slee and Kirwan, 2007; Zhou and Chen, 2007). Furthermore, following the research collaboration several articles have been published in on-line and printed magazines, in the DEIAgra working papers series (Canavari *et al.*, 2007; Chai *et al.*, 2008; Marchesini *et al.*, 2007), in international scientific journals (Canavari *et al.*, 2010; Huliyeti *et al.*, 2008), in national scientific journals (Huliyeti *et al.*, 2007; Liu *et al.*, 2007, 2008; Liu and Chen, 2007), and in an edited book published in 2009 by Wageningen Academic Publishers (Canavari *et al.*, 2009).

The evaluation of this part of the project indicated very positive impacts on scholars' learning, not just from an academic perspective but also from the experience of working in a different country, experiencing different cultures and different ways of working.

3. Conclusion

Clearly, interest and engagement with quality and organic food is growing in Asia as in Europe. However, there are important differences in attitudes towards organic and quality food and the development of the BEAN-QUORUM project has considerably advanced understanding of these issues. Sharing experiences through the network will be made possible by the use of Information and Communication Technologies, namely a web site and a social network platform (beanquorum. ning.com). Both will enhance communications among the team and other interested parties in industry and the wider support sectors.

Although the project is now closed, the foundation that had been created for the future may still play a relevant role in pooling resources and enhancing the value of members' research activities. The BEAN-QUORUM network is now established and new members are joining on a regular basis. This is the beginning of a consolidated network of researchers who will work with the business sector and NGOs to enhance European Asian understanding about organic and quality food. This is also an opportune time for developing a network such as BEAN-QUORUM, as the EU is actively interested in promoting the principles of geographical indications of origin for foods in third countries and the foundations that we have built will help further develop both European and Asian scientists' capacities.

The final outcome of the project, this book, collects some of the work performed during the project, eventually adapted to a teaching purpose. However, this book also builds on joint research activities that have been undertaken on the basis of external funding, since specific research activity was not funded by the BEAN-QUORUM project, and the main results of this collateral activity as well as the lessons learnt in terms of context specificities, methods, and approaches have been included in the case studies.

Finally, the book also contains papers prepared by scholars and experts who collaborated with the BEAN-QUORUM network and who gave a further contribution to the growth of the knowledge base built within the network. This is also a promising starting point for future collaboration within a growing network.

This book describes global trends in organic and quality food trade and connects them with recent developments in Asian and European market structures. Selected case studies illustrate the impact of organic and quality food production on topics ranging from sustainable rural development, to the potential of exotic new plant varieties to purchase decisions of European or Asian retail managers. Selected European markets are mirrored by the situation in Chinese and Thai markets. Finally, environmental issues concerning global trade of quality food are addressed.

The publication represents an attempt to provide a collection of meaningful studies which may be useful as a support to teaching marketing with a grasp on specific problems related to quality food products, not only in a European context, but also taking the perspective of an Asian marketer. We believe that this collection of papers, dealing with several topics and adopting many different approaches, represents a useful companion for a teaching course on food marketing at bachelor and master's level.

Acknowledgements

This publication has been produced with the financial assistance of the European Union. The contents of this publication are the sole responsibility of the editors and the authors of the single contributions and can under no circumstances be regarded as reflecting the position of the European Union.

References

Brunsø, K., Fjord, T.A. and Grunert, K.G., 2002. Consumers' food choice and quality perception (No. 77). University of Aarhus, Aarhus School of Business, The MAPP Centre, Aarhus, Denmark.

Canavari, M., Cantore, N., Castellini, A., Pignatti, E. and Spadoni, R. (eds.), 2009. International marketing and trade of quality food products. Wageningen Academic Publishers, Wageningen, the Netherlands.

Canavari, M., Centonze, R. and Nigro, G., 2007. Organic food marketing and distribution in the European Union (Techreport No. 7002). Alma Mater Studiorum University of Bologna, Department of Agricultural Economics and Engineering, Bologna, Italy.

Canavari, M., Lombardi, P. and Spadoni, R., 2010. Evaluation of the Potential Interest of Italian Retail Distribution Chains for Kamut-Based Products. Journal of Food Products Marketing, 16(1): 1-21.

Chai, J., Slee, B., Canavari, M., Chen, T. and Huliyeti, H., 2008. Study on the scope for reconstruction of the grazing livestock sector of Xinjiang based on organic farming methods (Techreport No. 8001). Alma Mater Studiorum University of Bologna, Department of Agricultural Economics and Engineering, Bologna, Italy.

Chai, J., Zhang, X., Zhang, J. and Chen, T., 2007. Study on Problems Faced by Xinjiang Organic Producers and Solutions. In: M. Canavari, D. Regazzi and R. Spadoni (eds.), International Marketing and International Trade of Quality Food Products. Proceedings CD-ROM of the 105th Seminar of the European Association of Agricultural Economists, Avenue media, Bologna, Italy: pp. 343-350.

Grunert, K.G., 2005. Food quality and safety: consumer perception and demand. European Review of Agricultural Economics 32(3): 369-391.

Huliyeti, H., Marchesini, S. and Canavari, M., 2007. The scenario of organic agriculture in Italy (in Chinese). Xinjiang Nongcun Jingji - Xinjiang Rural Economy 1: 51-53.

Huliyeti, H., Marchesini, S. and Canavari, M., 2008. Chinese distribution practitioners' attitudes towards Italian quality foods. Journal of Chinese Economic and Foreign Trade Studies 1(3): 214-231.

ISO (International Organization for Standardization), 2005. ISO 9000:2005 Quality management systems – Fundamentals and vocabulary. ISO, Geneva, Switzerland.

Liu, R. and Chen, T., 2007. Study on the Organic Agricultural Production Problems and the Mechanism Innovation in Yiwu-Xinjiang. Xin Xibu - New Western 9: 9-10.

Liu, R., Chen, T. and Haas, R., 2007. Experience and Significance of Organic Agriculture Development in Austria (in Chinese). Laoqu Jiangshe - The Construction of the Old Revolutionary Area 10: 62-63.

Liu, R., Chen, T. and Haas, R., 2008. Analysis of the Promotion of Organic Agriculture on Rural Economic Development - A Case Study in Yiwu, Xinjiang (in Chinese). Nongye Kaogu - Agricultural Archaeology 3.

Marchesini, S., Huliyeti, H. and Regazzi, D., 2007. Literature review on the perception of agro-foods quality cues in the international environment. In: M. Canavari, D. Regazzi and R. Spadoni (eds.), International Marketing and International Trade of Quality Food Products. Proceedings CD-ROM of the 105th Seminar of the European Association of Agricultural Economists. Avenue media, Bologna, Italy, pp. 729-738.

Marchesini, S., Huliyeti, H., Canavari, M. and Farneti, A., 2007. Attitudes towards Italian wine of practitioners in the Chinese distribution (Techreport No. 7003). Alma Mater Studiorum University of Bologna, Department of Agricultural Economics and Engineering, Bologna, Italy.

Morawetz, U.B., Wongprawmas, R. and Haas, R., 2007. Potential income gains for rural households in North Eastern Thailand through trade with organic products. In: M. Canavari, D. Regazzi and R. Spadoni (eds.), International Marketing and International Trade of Quality Food Products. Proceedings CD-ROM of the 105th Seminar of the European Association of Agricultural Economists. Avenue media, Bologna, Italy, pp. 111-125.

Oude Ophuis, P.A.M. and Van Trijp, H.C.M., 1995. Perceived quality: A market driven and consumer oriented approach. Food Quality and Preference 6(3): 177-183.

Slee, B. and Kirwan, J., 2007. Exploring hybridity in food supply chains. In: M. Canavari, D. Regazzi and R. Spadoni (eds.), International Marketing and International Trade of Quality Food Products. Proceedings CD-ROM of the 105th Seminar of the European Association of Agricultural Economists. Avenue media, Bologna, Italy, pp. 247-260.

Steenkamp, J.-B.E.M., 1990. Conceptual model of the quality perception process. Journal of Business Research 21(4): 309-333.

Zhou, L.-L. and Chen, T., 2007. Consumer Perception of Organic Food in Urumqi. In: M. Canavari, D. Regazzi and R. Spadoni (eds.), International Marketing and International Trade of Quality Food Products. Proceedings CD-ROM of the 105th Seminar of the European Association of Agricultural Economists. Avenue media, Bologna, Italy, pp. 173-186.

Part 1.
Looking east looking west

Organic food in the European Union: a marketing analysis

R. Haas[1], M. Canavari[2], S. Pöchtrager[1], R. Centonze[2] and G. Nigro[2]
[1]University of Natural Resources and Applied Life Sciences Vienna (BOKU), Institute of Marketing and Innovation, Feistmantelstrasse 4, 1180, Vienna, Austria
[2]Alma Mater Studiorum University of Bologna (UniBO), Department of Agricultural Economics and Engineering Viale Fanin 50, 40127 Bologna, Italy

Abstract

This paper discusses the European organic agricultural sector from a socio-economical point of view and from an EU perspective. The organic food and beverage market has been a niche market in Europe for many decades. But since the 1990s the number of organic farms has increased significantly in some European countries. In the last decade 'organic' has become a growth, and in some European countries, a mature food and beverage market due to ongoing governmental support, active marketing of national and international retail chains and constantly growing consumer demand. If organic food was originally the result of an ideological choice, anchored in an alternative culture, today it has become a more mainstream phenomenon. Organic now belongs to a specific lifestyle and to a finally acknowledged cultural model, which attracts human and financial resources on its own, producing profits and satisfying a steadily increasing market. This paper beginning with a description of the global organic food market analyses organic food and beverage marketing in Europe in respect to consumer segments and to the four marketing Ps (Product, Price, Place and Promotion).

1. Introduction

The international food standards, Codex Alimentarius, state: 'organic agriculture is a holistic production management system which promotes and enhances agro-ecosystem health, including biodiversity, biological cycles, and soil biological activity. It emphasises the use of management practices in preference to the use of off-farms inputs, taking into account that regional conditions require locally adapted systems. This is accomplished by using, where possible, agronomic, biological, and mechanical methods, as opposed to using synthetic materials, to fulfil any specific function within the system' (FAO, 1999).

The foundation of the development of organic products is in its recognition as a separate farming technique, with its own principles as compared to conventional agricultural production. The organic production certificate is given according to the European Union regulations, being Reg. 834/2007 related to organic agricultural and food productions and the following Commission Regulation (EC) No. 889/2008 where all levels of plant and animal production are regulated, from the cultivation of land and keeping of animals to the processing and distribution of organic foods and their control. Furthermore, Commission Regulation (EC) No. 1235/2008 of 8 December 2008 contains detailed rules concerning import of organic products from third countries.

The EU is placed second in the world concerning the area for organic agriculture, preceded only by Australia. The most important feature is that organic production shows an independent growth in comparison to the growth of conventional production. In fact, although the European land available for agriculture decreased by 2.5% from 1990 to 2000, thus going from 115.3 millions of hectares to 112.7, in the same period the land available for organic production increased by 1.2%. From this data, it is apparent that organic agriculture represents a significant economic opportunity for European farms and the development of rural areas. In many EU countries, growth in the organic sector offers new employment opportunities, not only at the primary sector level but also at the processing level as well as in the relevant services.

Another important aspect is the compatibility of the organic concept with another powerful rural development and marketing tool widely used in France, Italy, and other EU countries, such as the geographical indications (PGI/PDO) certification and labelling instruments. The compatibility of production rules is not always guaranteed, and the issue of overlapping controls is also of extreme importance (Canavari, 2007). In particular, such a certification system highlights either the distinguishing capacity of food products through protection labels such as traditional/local food; PDO and PGI or the peculiarities of the food production of each member country including traditions and organoleptic qualities (Lunati, 2006). Organic products fall in the category of 'value added' food (premium price) such as traditional products, because the consumer considers them as a speciality food.

The aim of this paper is to determine the role and the perspectives of organic food on the European agricultural and food market and to provide a quantitative description of the phenomenon of organic food both from the production and distribution side. Following the introduction a description of the collected data is given, before a general overview about the world wide organic production and market situation. The application of marketing mix instruments in the organic food value chain is described based on the example of selected European countries.

2. Materials and methods

This study called on different data sources (i.e. institutional and private bodies operating in the organic sector) either at the international (particularly European) or national levels.

Desktop research was built on the following bibliographic sources: IFOAM, BIOFACH, FIBL-survey, SOEL-durvey, EC 2005 Organic Farming in the European Union Facts and Figures, Commission General Direction Agriculture. This together with other sources at the national level: AgrarMarket Austria (AMA), Nielsen Austria/Germany, ISMEA/ACNIELSEN, BIO BANK, Italian Ministry of Agriculture (MIPAF), SINAB, Federbio, inspection and certification bodies.

The data collected (the reference period 2003-2007) describes in detail the organic food production in the EU. In particular, the organic market has been described with the following methodology:
- the organic agriculture surface area in the world;
- the number of organic farms in the world;
- the value of organic food consumer markets in selected areas (North America, Europe, Asia, Oceania, South America);
- the organic retail sales in EU-15 and EU-27;
- per capita expenditure for organic food in selected countries.

In this paper, the situation of organic food marketing in Europe with a particular focus on Italy and Austria, is also discussed. The analysis is structured according to the classical four marketing mix tools: product, price, distribution and communication. Finally, the authors draw conclusions from the previous analysis and identify trends of emerging markets and possible actions to support the growth of organic food.

3. General overview

Organic agriculture has gained a relevant market share on the supply side, but mostly in developed countries. In Australia a relevant surface area of mainly pasture and grazing land has recently been converted to organic agriculture. In Europe, quite a large share of agricultural surface area is now managed according to the principles of organic agriculture. Organic agriculture surface area in the world accounted for more than 32 million hectares in 2007, with 5% growth compared to 2006. The number of producers in 2007 was around 1.2 million. (Willer *et al.*, 2009: 26). The organic area increased by 1.5 million hectares between 2006 and 2007 and covers 0.8% of agricultural land worldwide. 'The regions with most organically managed land are Oceania, Europe and Latin America. Australia, Argentina and Brazil are the countries with the largest organically managed land areas, the highest shares of organically managed land are in Liechtenstein, Austria and Switzerland. Almost half of the world's organic producers are in Africa. The countries with the highest numbers of producers are Uganda, India and Ethiopia' (Willer *et al.*, 2009: 26). Australia has about 37% of the world's organic agriculture surface, followed by Europe (24%) and Latin America (20%) (Figure 1).

In Europe about 7.8 million hectares had been assessed as organic farming land at the end of 2007. The list of EU member countries is lead by Italy with a surface corresponding to 1,150,253 hectares, followed by Spain (988,323 ha), Germany (865,336 ha), UK (660,200 ha), France (557,133 ha) and Austria (372,026 ha). The biggest countries in respect to organic farming land in Asia are China (1,553,000 ha), followed by India (1,030,311 ha) and Indonesia (66,184 ha) (Willer and Kilcher, 2009: 279). In 2005 the USA had 1,640,000 ha under organic management (USDA, 2007).

The number of organic farms is constantly increasing worldwide. It is difficult to obtain precise numbers of organic farms due to methodological issues, for example some countries report the number of producers per crop and therefore double counting occurs.

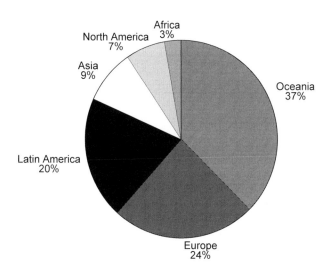

Figure 1. The organic agriculture surface area in the world (year 2007) (FIBL/IFOAM Survey, Willer et al., 2009: 27).

In the EU area, the farms producing organically are about 210,000 of which: 45,231 in Italy, 23,769 in Greece, 19,997 in Austria, 18,703 in Germany, 18,226 in Spain and 11,978 in France; the above listed countries account for about 65% of the farms (Willer and Kilcher, 2009: 285). In comparison to these numbers the Asian country with the most producers is India with 195,741 (Willer and Kilcher, 2009: 285) and in the USA 8,493 producers (USDA, 2007) were reported (Figure 2).

The global market of organic products is constantly growing meanwhile the gap between producers and consumers is widening. The strongest growth rates in organic production are in developing countries, which export most of their national organic production. The organic managed area in Africa, Asia and Latin America shows three digit growth rates since the year 2000 compared to other regions of the world with two digit growth rates (Ratanawaraha *et al.*, 2007: 20). The demand is concentrated in wealthy developed countries such as the EU-27 or North America. The EU-27 and North America cover around 97% of the demand for organic food (Sahota, 2009: 63). The global gap between demand and supply of organic production is a challenge for the whole organic sector. Keywords such as carbon footprint, climate change, and long transport routes show that the organic market is changing. Since its European origin, organic food has been seen as a local or regional food and still in some European countries is perceived as local/regional food (see Haas *et al.*, 2010). However due to arising demand, the image of organic food as local/regional food is less and less in accordance with the market reality.

In 2007 the world total turnover of organic food & drinks was estimated up to 46 billion US$; with organic sales in Europe of 25 billion US$ and 18 billion US$ in the USA (Sahota, 2009). Interestingly the size of organic sales in the USA is around 70% of the sales in Europe but the farmland under organic management in the USA is only around 21% of the EU organic farmland and the number of US organic producers is only 4% of the number of EU organic farms (own calculations based on numbers given above). So while Europe is producing a significant share of its organic demand, the USA has to import it from countries such as Mexico, Brazil or Chile (Sahota, 2009: 60).

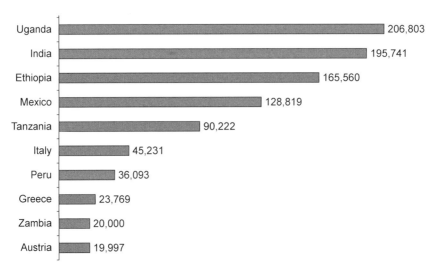

Figure 2. Number of organic farms in the world (year 2007) (Willer and Kilcher 2009: 283).

The global market has tripled in value since 1999 and in 2007 its growth was 5 billion US$ a year (Sahota, 2009: 59). Nevertheless due to the global economic crisis growth rates are slowing down, traditional organic consumer segments will stay loyal but it can be expected that the market segment of impulsive shoppers is switching to less expensive conventional food & drink products.

The year 2007 was dominated by supply shortages, high fuel costs and an increased share of farmland under biofuel crops, leading to high food & drink prices and increased inflation rates. The financial crisis brought a sudden end to this development and the stifled demand could lead to oversupply of organic products in categories such as meat, dairy, fruits and grains in the near future. Producers in developing countries will be strongly influenced by this new market situation forcing them to develop their national markets to be able to cope with the decrease in international demand (Sahota, 2009: 63).

As mentioned before, Europe represents the biggest and most mature market for organic food & drink in the world, accounting for 54% of all global sales (Sahota, 2009: 60). Table 1 shows the value of the organic food markets for all continents and the main organic food countries in Europe and North America. Africa, Asia, South America and Oceania cover around 3% of the organic food market. Again these numbers emphasize besides the importance of these countries as producers their negligible role on the demand side on the global dimension.

Europe is a very diversified market, encompassing highly sophisticated organic markets in countries such as Germany, the United Kingdom, France or Austria on one side and on the other side, countries which are just beginning to 'discover the possibilities that are arising from the global organic boom' mainly located in Southern and Eastern Europe and in some parts in Northern Europe (Van Osch *et al.*, 2008: 376).

Table 1. The value of organic food consumer markets in selected areas (2007).

Area	Country	Value (in millions of euro or US$)	
North America			20,000 (US$) [1]
	USA	18,000 (US$) [1]	
	Canada and Mexico	2,000 (US$) [1]	
Europe			25,000 (US$) [1]
	Germany	5,300 (€) [2]	
	United Kingdom	3,080 (€) [2]	
	France	2,100 (€) [2]	
	Italy	1,595 (€) [2]	
	Switzerland	783 (€) [2]	
	Austria	739 (€) [2]	
Asia			576 (€) [3]
Oceania			288 (€) [3]
South America			120 (€) [3]

[1] From Sahota (2009: 59).
[2] from Van Osch *et al.* (2008: 376).
[3] from Biofach 2005 and Fibl Survey.

The per capita consumption in Europe illustrates the diverse market situation of the European countries. Switzerland, a non-EU country, is the leader in Europe concerning per capita expenditure, which is higher than 100 euro per year. Austria follows second with 89 euro, Luxembourg third with 86 euro per capita spending per year. Italy ranks 9th, with an average yearly expenditure of 25 euro (Figure 3). The forecast of per capita spending reflects the slowing growth rates for most of the countries except Denmark.

4. The European consumer for organic food & drink

Before the European organic market is described in respect to the marketing mix, it is necessary to describe consumers of organic food. As in many consumer markets it makes no sense to talk about the typical average organic consumer, but there are specific consumer segments, which exist in more or less similar shares in many European countries.

Based on lifestyle and value segmentation an important consumer group for organic products in Western Europe is LOHAS (Lifestyle of Health and Sustainability). After the dominant hedonistic consumer orientation in the 1970's and 1980's the LOHAS segment represents consumers, who are neither willing to accept lower quality or bad taste for organic products, as had been the case in the early 'idealistic' stages of the organic market. Nor are these consumers willing to accept products, which are produced under environmental or socially harmful conditions. These consumers look for products that are healthy, and produced in an environmentally friendly and socially responsible way. Therefore organic and fair trade often goes hand in hand. In Austria for instance fair trade coffee and fair trade bananas show annual growth rates of around 30% for the last three years with a significant share of organic fair trade coffee and bananas (Holley-Spiess and Möchel, 2009: 6).

In recent years a shift towards older consumer segments has been observed (ISOE, 2003, Nielsen, 2006). In Germany in the age groups of 40-49, 50-59 and 60-69, the share of organic food shoppers

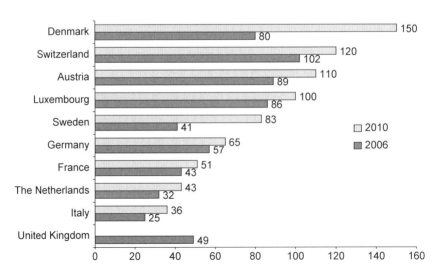

Figure 3. Per capita expenditure (in euros) for organic food in Europe (year 2006 and forecast 2010) (Van Osch et al., 2008). 2010 forecast for UK not available.

is significantly above average (ISOE, 2003: 8). Categorized in life cycle segments, these belong mainly to the 'empty nesters' and 'seniors', (Nielsen, 2006). Empty nesters and seniors are also the segments, with the highest percentage stating that they are searching for food products with healthy added value (Nielsen, 2006). ISOE (2003) found in these segments also a higher share of women.

The rationale behind it could be that empty nesters have a higher share of their income available for food after their children have left home, and/or that they (together with seniors) are more health conscious due to the fact that they are feeling the first signs of aging. The 50+ segment responds, in an open question, that the most important trigger for shopping of organic food is illness or health concerns (AMA, 2007).

ISOE identified in a quantitative representative sample, for Germany five consumer segments for organic food. In respect to purchase frequency and number of purchased products their biggest consumers segment is the 'holistically convinced'-segment with a share of 42% of the organic food shoppers in Germany. Again empty nesters in this segment are above average. Based on their lifestyle and value orientation this segment is comparable to LOHAS, looking for quality, environmental and social justice, sensuality (taste) and spirituality (ISOE, 2003: 10).

Overall health is the main motive for the purchase of organic food, also on a global scale (based on an internet survey of Nielsen with more than 21,000 consumers; Nielsen, 2006), health for oneself and health for ones children. In respect to the importance of health as a motive, one could call organic food a form of 'soft functional food'. The ranking and importance of other motives are varying in respect to country or culture. Other motives are taste, environmental friendliness and animal welfare (Zanoli *et al.*, 2004). For example the motive 'better for the environment' is significantly more important for European consumers than for North American, Latin American or Asian Consumers; or for e.g. the motive 'animal welfare' is much more important for German consumers than for other Europeans (Nielsen, 2006). Age seems to have an influence on the 'nature' of the health motive. Young consumers have more an enjoy-wellness-health orientation, parents connect health with responsibility-security and for older people a health sustaining view is dominant (Spiller *et al.*, 2004).

A means end chain analysis with Austrian consumers regarding their motive structure, revealed that after the health motive, the second most important motive is a hedonistic value represented by statements such as 'enjoy my life', 'delight', 'to do myself something good', which is strongly connected to 'better taste' as a consequence of eating organic and better taste is connected with the specific product attributes 'freshness', 'naturalness' and 'whole food' (AMA, 2007).

In respect to socio demographic variables (wealthy) 2 person households or households with 3+ are represented above average in German and Austrian surveys (AMA, 2007; ISOE, 2003). The first openly mentioned trigger of Austrian families with children, for buying organic food is 'responsibility for my children' (AMA, 2007). ISOE (2003) called this segment the 'wealthy and demanding ones' and estimated a share of 23% of the German organic food shoppers. Overall there are egoistic motives (health, taste) and altruistic motives (environmental protection, animal welfare). Especially for light-users and not-regular consumers of organic food, egoistic motives are much more important purchase motives than altruistic ones. Applying the Maslow's hierarchy of human needs, it is easy to see that first comes safety and security needs before altruistic motives are applied (Fricke, 1996: 167).

A spontaneous association test about organic food with Austrian consumers revealed that they associate words (order based on the counts of statements) such as 'no chemistry' (related semantic statements: no pesticides, free of residues, no artificial fertilizers,…), 'healthy nutrition' (no genetic engineering, fruits, vegetables, …), and in third place 'Austrian origin' (support small farmers, buy from our farmers, traceability, …), before they mentioned words such as 'nature, environment', 'animal welfare' or 'quality, taste' (AMA, 2007). These results of the association test, especially concerning the attribute 'origin' may also be true for other European countries. A survey with around 400 experts from the German food retail and industry sector underlines the importance of country of origin. Asked if organic products should be regional and should have short transport routes around 80% of the experts agreed (Anon, 2008: 55). Nevertheless one has to be cautious by interpreting these results because 'region' is a somehow fuzzy term. Haas *et al.* (2010) found in their study differing answers concerning the area of a region. In the case of Austrian owners of specialised food stores a region for example is a specific sub-area of Austria for e.g. Tyrol (which is a province of Austria) or parts of Tyrol. So region for them is recognised as having a distinctive identity on the basis of its social and/or economic and/or natural characteristics. This could be seen as a confirmation of the possible link and market overlapping between organic and local/traditional products (PDOs and PGIs). On the other hand Austrian purchase managers in the retail sector mentioned that Austria as a whole is a region or a combination of Austria and parts of neighbouring countries. Looking at these answers leads to the hypothesis that the more managers/entrepreneurs are concerned about achieving economies of scale the bigger they define the geographic area of a region.

The main barrier for the purchase of organic food is price (AMA, 2007; ISOE, 2003; Nielsen, 2006). Non-shoppers and light-users of organic food especially mention this barrier; their number is significantly higher then heavy-users of organic food. The second most mentioned barrier is credibility (AMA, 2007; ISOE, 2003; Nielsen, 2006). Statements such as 'I doubt that organic food is sufficiently controlled' (ISOE, 2003: 8) or 'Organic on the label, doesn't mean organic in the product' (AMA, 2007) are typical for the lack of credibility.

5. Marketing mix aspects

In the following sections we discuss the situation of organic food marketing in the European Union member countries with a particular focus on Italy and Austria, with anecdotal information about other selected European countries. The analysis considers the 4 marketing mix tools:
- Product: EU regulations, certification and market shares.
- Price: premium price, price differences in market channels.
- Distribution: large retail, traditional retail, specialized shops, food service channel.
- Communication: advertising, promotion, public relation activities of private/public institutions.

A fifth tool considered is branding (branding policy, private labels versus producer brands, …), usually seen as a product attribute, but here seen as an indicator for the maturity of the organic food markets on national and international markets.

5.1 Product

In June 2007 the European Commission passed the Regulation EC 834/2007, which was enacted at the beginning of 2009. This regulation repeals the Regulation EC 2092/91. The Regulation EC 834/2007 represents the basis of the EU regulation on organic farming for 'plant-based agricultural, not transformed products'. It regulates: labelling, standard production methods, control system, provisions relevant to organic food import from non-EU countries, products for the soil manuring and amending, phytosanitary products admitted for organic pest management. Further the Reg. EC 889/2008 lays down detailed rules for the implementation of 834/2007 with regard to organic production, labelling and control. Arrangements for imports of organic products from third countries are regulated in Reg. EC 1235/2008.

As previously mentioned, a lack of credibility is one major barrier for consumers to buy organic food. Therefore a transparent and trustworthy certification system is of paramount importance. Any farm aiming at producing or processing organic products has to be certified by a control body guaranteeing the compliance with regulation EC 834/2007. For instance, every European country has to authorize control bodies, for e.g. Italy has authorized 16 control bodies to certify Italian organic products; Austria has eight certified control bodies. Based on the EU regulation control bodies have to comply with the EN norm 45011. Each control body is identifiable over a unique code. A complete list of all accredited EU control bodies has to be published by the EU commission due to EC 834/2007. For e.g. the code 'AT-T-01-Bio' leads to a control body in Tyrol Austria (AT standing for Austria, T for Tyrol).

Not only the production process, but also the product itself has to be certified. Each step in the value chain has to be certified on special registers held by the specific national agricultural ministries. The average cost of certification in Italy for an agricultural farm with 3-4 hectares varies from 300 to 400 euro per year, depending on the region, the production system and on the controlling organisation. In Austria an organic farm has on average about 20 hectares, the costs of certification are about 250 euro per year, depending on the controlling organisation.

The Regulation EC 834/2007 clarifies the discipline from the point of view of both consumers and farmers. From 1 July 2010 all organic pre-packaged food, which is produced or on sale within the EC, has to be labelled with the Community organic production logo. The design of the Community logo has to follow the model in Regulation EC 889/2008, which includes the term 'organic farming' in the language of the respective country. When the Community logo is used, an indication of the place where the agricultural raw materials of the product are farmed, should also appear in the same visual field as the logo. After 1 July 2010 the use of the Community logo must be combined with an indication of the origin where the agricultural raw materials were farmed. It should be mentioned that the raw materials can originate from 'EU Agriculture', 'non-EU Agriculture' or 'EU/non-EU Agriculture'. In the case that all raw materials have been farmed in only one country, the name of this specific country, in or outside the EU, can be mentioned instead. National and private logos may be used in the labelling, presentation and advertising of products, which satisfy the requirements set under this regulation.

In order to be labelled as 'organic' the end product must contain at least 95% of organic ingredients; if the end product contains less than 95% of organic ingredients it may not be named organic, but the ingredients with references to the organic production may appear in the list of ingredients. Products containing Genetically Modified Organisms (GMO) will not be labelled as organic food (a maximum percentage of GM not higher than 0.9% for each ingredient, due to accidental contamination, is allowed); imported organic food will be admitted only if it complies with EU rules or if they have equivalent guarantees. It is mandatory to label the product with the code number of the authorized control body to which the operator is subject.

As far as the main types of organic food produced in Europe are concerned, they can be divided into basic organic food (fresh and processed) and value-added organic food. In the highly developed organic national markets of the EU almost every conventional food is offered as organic food. For e.g. the Austrian private label Ja natürlich (Yes naturally) has an assortment of 1000 different products (SKU, Store keeping units) ranging from fresh fruits, sausages to convenience or frozen food. To give a 'feeling' about market shares in mature organic markets Figure 4 and 5 represent data from Austria based on two different panels: a retail panel and a household panel.

Figure 4 shows the market shares in sales in Austrian retail, including the discounter chains Hofer and Lidl and also drugstore chains such as DM or Schlecker. The Nielsen MarketTrack is a retail panel covering shopping data from conventional retail and drugstore chains; these numbers stem from scanner data or are based on estimations from collected bills. Overall organic food has a market share of 4.4%, which is still a small percentage in respect to the fact that the Austrian food market can be considered to be a mature organic market (The household panel of AgrarMarkt Austria reports a market share of 5.1% in value and around 6% in volume; AMA, 2009). Looking at the strongest product categories such as baby food (39%) and health foods (26.2%), the shopping preferences of organic customer segments from Section 4 (families, empty nesters and seniors) can be seen again. Obviously families buy baby food, while it can be assumed that mainly empty nesters and seniors buy health food (whole food) due to their growing interest in a healthy lifestyle.

Figure 5 gives a more detailed picture about market shares (volume) of dairy and fresh products based on Austrian household shopping data. Organic fresh milk and extended shelf life milk (ESL) have a market share of 12.7% (volume) of all fresh and extended shelf life milk in Austria. Organic eggs in second place cover a share of around 10% of all egg sales. Obviously the categories 'organic meat & chicken' and 'organic sausage & ham' are of minor importance in respect to conventional meat and sausage sales.

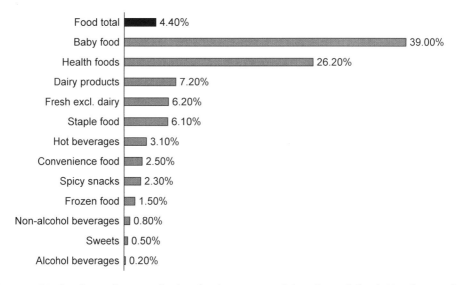

Figure 4. Market share of organic food in the Austrian retail (retail panel data). Numbers are based on sales in value (Nielsen, 2007).

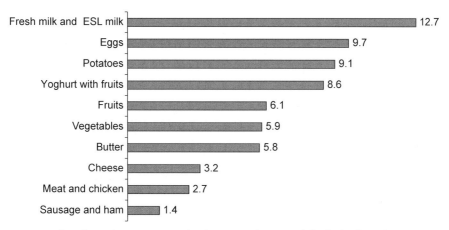

Figure 5. Market share (in percentages) of organic dairy and fresh food in the Austrian retail, n=2,500 households (Austrian household panel data) (AMA, 2009).

5.2 Brands, trademarks and private labels

The latest European food scandals (BSE, dioxin, etc.) have partly contributed to the development of the organic food market. Moreover in the last years, the improvement of distribution channels has played an important role in the promotion of purchases.

Indeed, the entry of large retail companies in such markets and the creation of important groups, improved the marketing and the distribution of organic foods. In fact, one of the most important elements for the spreading of organic foods is given by marketing policies and advertising realised by public branding (EU, member states, regional authorities), collective branding (associations, consortia) and private branding (retailer's brand, producer's brand, certification body's brand).

Differentiated policies have the advantage to meet the national consumers needs and preferences. But many companies, NGO's or officials confuse the concept of a brand with a seal or trademark. There are many organic seals on the market, for e.g. from farmers associations or national labels. The variety and number of which, is mostly only contributing to confuse the consumers. Most of the seals are far from being considered a brand. In many cases they can be compared to a trademark. A trademark is a word, name, symbol, colour, scent or sound used in trade to distinguish goods or services and a trademark is a legal construct designed to protect consumers from confusion. A brand has much more intangible assets. Therefore marketers say a brand has an identity, a brand is a promise and a brand stands for specific values, culture and personality. Obviously the value of the organic brand depends on its reputation, reliability and credibility.

There is a European organic seal available, but there are differences among EU countries: in Germany, France and Austria national seals have been created. Any EU and non-EU organic company may request and obtain the right to use it if they comply with the rules:
• The Netherlands, Belgium, Sweden and Italy have no national seal but several control bodies authorised by the Ministry of Agriculture.
• Denmark has established a national seal for organic foods 'Ø' (*Økologimaerke*) controlled by the State.

- In Switzerland (non-EU country) the most widespread seal is the private bio-mark Biosuisse, any EU organic company producing in conformity with Swiss rules may require the brand.
- France has a national logo for organic products – the AB-Logo (AB=*Agriculture Biologique*) – which is owned by the French state.

In Germany there are currently eight organic producer organisations. They contain about 60% of Germany's organic farmers. The organic producers' organisations all own legally protected seals with which certified farms and certified processors can be labelled. These seals are familiar to German consumers, especially those of Demeter (which is actually a Biodynamic agricultural brand), Bioland and Naturland.

In Austria, in 1994 the Ministry of Agriculture and Forestry introduced a label to guarantee product safety to the consumer. The *Austria Bio-Zeichen* ('Austria Organic Label') may be used by approved farmers, processors and trading companies. It guarantees that the food bearing this label originates from organic farming. In addition, if the label is coloured red and white, it certifies that at least 70% of the ingredients originate from domestic organic farming. If the Austrian label is coloured black and white it is a supplementary logo for non-domestic organic products. A new trend is represented by the national labels, which also include the local origin of the product (for example Hessen, Baden Württemberg, or Bayern). These labels respond to the preference for local, regional organic food.

Concerning branding it has to be emphasized that private labels owned by retailers dominate the organic market. All major European retailers have launched organic private labels (Table 2). Interestingly in 2005 Carrefour has subsumed 'Carrefour Bio' under a new Carrefour umbrella brand with the name 'Carrefour Agir' ('agir' means 'to act') and now has four subbrands:
- Carrefour Agir Bio: organic food products;
- Carrefour Agir Eco Planète: environmentally-friendly;
- Carrefour Agir Solidaire: fair trade products;
- Carrefour Agir Nutrition: health foods (Carrefour, 2009).

Experts see this move of Carrefour as a sign for a reduced engagement in the organic brand. Nevertheless Carrefour was one of the first in France to collaborate strongly with organic farmers and focused on coordinating all steps in the organic value chain. REWE, the German retailer, who owns the strongest Austrian private label 'Ja natürlich' (Yes naturally), also cooperates strongly

Table 2. Retailers private labels in Italy (year 2004) (Bio Bank, 2004).

Retailer	Label	Year of launch	References (number)
Carrefour	ScelgoBio	2000	221
Conad	Conad - Nuovi prodotti da agricoltura biologica	2000	50
Coop	Bio - Logici Coop	2000	307
Crai	Crai Bio	2001	39
Despar	Bio, Logico	2001	80
Esselunga	Esselunga Bio	1999	500
Gruppo pam	BioPiù	2000	42
Rewe italia	Si! Naturalmente	2001	160
Selex	Bio Selex	2001	17

with farmers and processors along the supply chain. Since they started in 1994, they showed a strong commitment in developing organic products together with farmers and processors.

There is a unique situation in the innovation process in the conventional food sector compared with the development of new products and brands in the organic food market. In the latter innovation was predominantly initiated by the food retailers rather than the food processing industry. This may have several reasons. One may reside in the atomistic and polypolistic small-scaled family based structure of organic agriculture. Even bigger organic farmer associations, lack the capital and in many cases the marketing know-how to create a brand. The food processing companies had obviously no interest in dealing with a heterogeneous small-scaled supply market and the associated extra switching cost due to the necessity of separate storage and processing facilities. Only the retail sector was willing to work with farmers and processors to develop organic private labels because they understood that organic products are a unique opportunity, a strategic necessity in a globalized anonymous food market. All retail chains in Europe have suffered under food scandals and an increasing share of replaceable national and global food brands, (those, which the consumer can buy at any supermarket independent of the supermarket chain). For European consumers buying food also means to make a statement about their identity, and in some way regional organic food gives identity back to the consumer. By offering distinct organic private labels the retailers profited from an image transfer, by offering food with an environmental friendly, healthy and regional image.

The 'bio-logici Coop' is the most mentioned private label Coop by the Italian consumers, in comparison with the 'eco-logici Coop' (environmental friendly) and 'solidal Coop' (fair trade) brands. In many European countries organic private labels have shown increasing sales in the recent years. One of the main reasons for this growth is due to the fact that private labels reduce the gap between the price of organic food and that of conventional products, which is generally quite high (from 30% to beyond 100%). In Austria the strongest organic brand is the private label 'Ja natürlich'. Introduced in 1994 'Ja natürlich' had 2008 a turnover of 272 million euro in Austria, which represented 45% of total organic *retail* sales in Austria and around 30% of total organic food sales over all distribution channels in 2008 (Holley-Spiess and Möchel, 2009; Poschacher, 2009). The product range of 'Ja natürlich' covers more than 1000 store keeping units (SKU), the most successful categories being bread and bakery (+7% growth), fruit and vegetables (+10%), dairy products and cheese (+11%). To get a feeling for the strength of 'Ja natürlich' it is helpful to compare it with one of the strongest organic brands in the USA, the private label called 'O' Organics' from the retailer Safeway, which reported sales of 300 million US$ in 2007 (Sahota, 2009)!

5.3 Price

One of the controversial issues limiting the development of the organic market is the price level for organic products. Price comparisons of different countries without including the costs of living and the general economic development such as GDP, inflation rate or unemployment could lead to wrong conclusions. Nevertheless this section uses price levels of selected European countries to illustrate at least the main price issues.

As a general rule, organic products receive a higher price than conventional products, but prices diverge depending on the country and on the product (on average more than 30%). The price premium is justified because of higher input costs and lower yields on the farm level and by higher quality due to higher environmental friendliness of the production. For example, Figure 6 shows the price premium at the retail level in Austria for organic milk, butter and cheese compared to

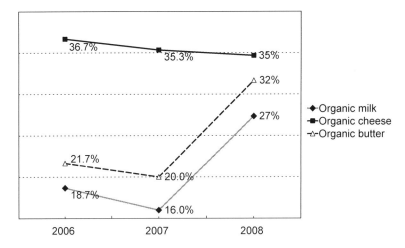

Figure 6. Price premium for organic dairy product categories in Austria (AMA, 2009). Average prices over all retail chains.

their conventional counterparts. Organic cheese has a price premium around 35% plus, while organic butter and milk show strong price fluctuations over the three years, reaching for milk for e.g. a price premium spanning from 19% to 27%. According to the report of Ismea/AcNielsen for Italy (2005) the price difference between organic food and conventional food in 2004 were 35.9% for the wholesale phase and 61.2% for the retail tier of the organic supply chain. These differences in price premia on the retail level between Austria and Italy have to be interpreted from the point of view that the Austrian organic market is more mature and shows higher levels of competition, especially from discount retailers, which puts more pressure on the price margins.

For organic apples, farm gate prices resulted to be the lowest in Italy (0.45 euro/kg) and comparable with the prices of conventional apples; about 60% of the EU-15 organic apples sales originated from Italy (Willer and Yussefi, 2005). On the opposite, farm prices for organic apples were the highest in Denmark (1.48 euro/kg) and in the United Kingdom (1.42 euro/kg), which were both net importers of organic apples. Farm price premia for apples were high in most EU member States except in Italy (2%), France (56%) and Austria (67%). This data is based on a survey conducted in 2001 for farm prices and in 2002 for consumer prices. The collected farmer prices are average prices, which farmers received when they sold their products to wholesalers or processors. The consumer prices were collected in different types of shops, selected according to the relative importance of sales channels in each country (Schmid *et al.*, 2004). These reduced farm price premia reflect the basic market law, since supply of organic apples in these countries was higher than demand. Italy could only sell 80%, Austria 85% and France 95% of its organic apples as organic, the rest had to be sold as conventional apples (Schmid *et al.*, 2004). Organic consumer prices for apples varied from 2.41 euro/kg in Italy to 3.65 euro/kg in the United Kingdom. Price *premia* for organic apples ranged extremely, from 37% in Sweden to 283% in Portugal. Table 3 shows price premia for selected organic products in several European countries at the farm gate level and at the retail level. The big variance in price premia illustrates the heterogeneity of the European organic food market. It seems that national particularities still have a strong influence on price building mechanisms.

Table 3. Farm-gate and consumer prices for selected products and countries (€/kg, year 2005) (EC, 2005).

	Farm-gate price premium				Consumer price premium			
	Min		Max		Min		Max	
Wheat	IT	19%	NL	189%	IE	29%	PT	203%
Apples	IT	2%	DK	333%	SE	37%	PT	283%
Potatoes	SE	71%	IT	293%	IE	30%	GR	170%
Pork	DE/AT	45%	NL	132%	PT	0%	GR	165%
Beef	DK	17%	ES	190%	PT	4%	LU	126%
Eggs	AT	25%	GR	329%	DK	17%	GR	231%
Milk	DK	19%	GR	129%	AT	6%	ES	58%

AT: Austria, DK: Denmark, ES: Spain, GR: Greece, IE: Ireland, IT: Italy, LU: Luxembourg, NL: the Netherlands, PT: Portugal, SE: Sweden.

In particular, Ismea/AcNielsen (2005) carried out a study on five organic fruit and vegetable products (oranges, potatoes, carrots, salad, and tomatoes) in order to analyse the price formation along the supply-chain in Italy. Comparing the price levels from farm to the consumer, during the first semester of 2004, the price of the five examined products increased on average by 125%, with peaks of up to 155% for tomatoes and 147% for carrots, while the lower margin was recorded for the salads (95%). Wholesalers on average applied a price margin of 13.75% (varying between 30% for tomatoes and 3.7% for oranges). Retailers applied on average a 100% price increase (ranging between 69.7% for the salad and 138% for oranges (Figure 7).

A further analysis studied packed organic foods including: cereals, milk producing products, olive oil, fruits and vegetables. Only 5 products have been analysed in this study on the specialised shops because of a lack of data: eggs, yoghurt, super-fine pasta, extra virgin olive oil and fresh

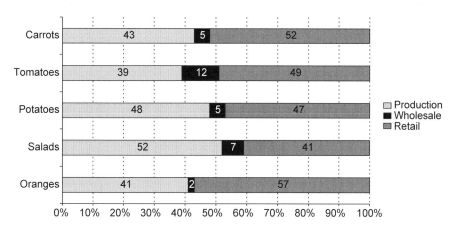

Figure 7. Price margins along the organic fruit and vegetable supply chain in Italy (year 2004) (Ismea/ACNielsen, 2005).

milk. By indexing the percentage variations, and considering the average corresponding to 100, it becomes clear that higher prices than the average are applied in the specialised shop (+31,7%), in traditional shops (+3,2%), in large retail (+0,3%), while in discount market and convenience stores prices are lower (respectively -8,3% and -16,1%) (Figure 8).

In specialised shops, higher prices are applied to specialty pasta (+52%) and olive oil (+52%), while in traditional retail higher prices are applied to biscuits (+42%). In discount markets lower prices are recorded for soybean beverages (-44%), olive oil (-20%) and yoghurt (-17%), while in convenience stores for fresh vegetables (-38%) and biscuits (-32%).

A comparison of price differences between conventional supermarkets and specialised organic stores in Germany, France and United Kingdom lead to similar results (even though the sampled prices reflect only a snapshot for the period January/February 2008). In general the price levels in specialised shops are higher than in conventional supermarkets. Figure 9 shows the difference in prices (plus or minus) in specialised organic shops compared to conventional supermarkets. Only eggs in France and United Kingdom were cheaper in specialised shops, all other sampled products were more expensive, the highest price difference was reported for olive oil with 109% in the United Kingdom.

5.4 Distribution

In the first part of this section, we analyse market trends and strategies adopted by large retailers and specialised shops in selected European countries; while in the second part we examine out-of-home channels distributing organic foods.

Distribution is a key factor for the growth of the organic sector. In almost all 27 European countries conventional retail chains have played and still play a crucial role in developing the organic market. The retail sector has the most influencing power to coordinate the manufacturers and producers, especially for its private labels, but also for the supply of other organic brands. Offering organic products on the retail level results, in most cases in sales of significant volumes

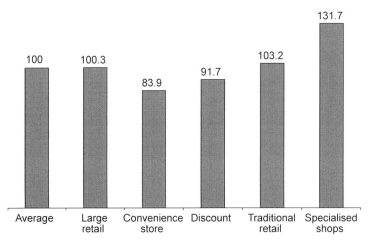

Figure 8. Comparison of retail prices (index numbers) for different channels in Italy (year 2004) (Ismea/ACNielsen, 2005).

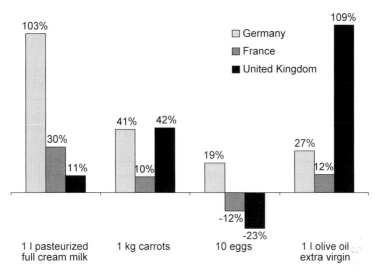

Figure 9. Price differences of specialised organic stores compared to conventional supermarkets in selected European countries (Van Osch et al., 2008; own calculation).

leading to profitable economies of scale. The decision of retail discounters in recent years to offer organic food increased the volume of organic food sold, but also increased the competition for the specialised organic retail sector, especially in the mature European organic markets (Van Osch, 2008: 382).

The specialised shops face tremendous competition all over Europe, forcing them to focus on their core competencies and focus on niche strategies. On one side the conventional retail sector is constantly growing and on the other side, new players – organic supermarkets – have entered the market. The opening of specialised large supermarkets offering only organic products is an emerging trend in many EU countries. In Europe the global trend of organic supermarkets is composed of the following organizations: NaturaSi, Planet Organic, Ecoveritas, Biocoop, La Vie Claire, Ekoplaza, Vierlinden (Table 4).

The marketing strategies adopted by specialised food shops are different from those adopted by large retailers. Specialised food shops below 150 m^2 face significant problems to achieve sustainable growth. They also face the challenge of assortment width compared to assortment depths. For example, in Austria from a survey of 140 specialised shops only 43% offered a full assortment (Kreuzer, 2009). These shops follow a niche strategy by offering products for specific target groups such as vegans or people with food allergies. On the other side both a broad and deep assortment is the advantage of organic or conventional supermarkets. REWE Austria offers more than 1000 and SPAR Austria offers 650 organic SKUs (these two are the biggest retailers in Austria with a market share of 59% approximately).

In particular, the specialised food shops rely upon the following factors: (1) a specialised product assortment focusing on the preferences of their loyal regular customers; (2) regional products at lower prices because of direct purchases from producers; (3) higher prices for national or international products, due to lower economies of scale and higher administrative costs; (4) a direct relationship with the consumer; (5) skilled staff offering advice; (6) additional services

Table 4. The global trend of European organic supermarkets (www.organic-services.com; Kreuzer, 2009).

Organic supermarkets in Europe	
NaturaSi (Italy)	40 outlets in Northern and Middle Italy and 1 outlet in Spain
Planet Organic (United Kingdom)	2 outlets in London
Ecoveritas (Spain)	7 outlets: Barcelona (5), Granollers (1), Andorra (1)
Biocoop (France)	104 supermarkets (223 outlets in total)
La Vie Claire (France)	20 supermarkets (120 outlets in total)
Ekoplaza (Netherlands)	1 outlet and 15 planned
Vierlinden (Germany)	1st outlet in Dusseldorf and 4 planned full product range
Basic, Bio Maran, Biomarket, Denn's, Mayreder's Bio Discount (Austria)	21 outlets

to customers for e.g. warm meals for lunch; (7) an ability to awake the interest of customers on ethical and environment-related topics and (8) more private labels than producer brands.

The large retailers instead, rely upon other factors: (1) the possibility to buy organic and conventional products at a single point of sale; (2) lower prices (but high premium and easier comparison with conventional food); (3) presence of specialised organic private labels on the shelves (to stimulate and promote organic foods); (4) organic food offered as a strategic means to improve the image of the retail brand.

Van Osch *et al.* (2008, 378) analysed 27 European countries (incl. Switzerland, Croatia and Norway; excl. the Baltic states) and found that in 15 of the 27 countries more than 50% of all organic food & drink is sold through the distribution channel 'conventional supermarket' (Figure 10). The highest market share for conventional supermarkets is in Northern European countries. In Sweden, Finland, Norway and Denmark consumers mainly buy organic produce

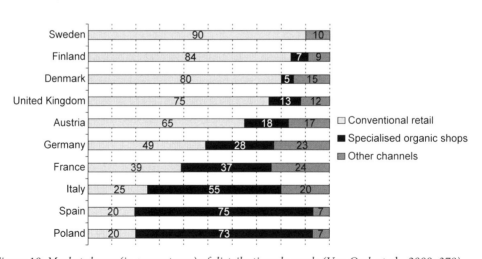

Figure 10. Market shares (in percentages) of distribution channels (Van Osch et al., 2008: 378).

from conventional supermarkets. In countries such as Poland, Spain, Italy, Greece or Portugal specialised organic shops are still the dominant players covering at least 50% of all organic sales.

An increasing share of developing countries is exporting to Europe. In fact, the number of producer co-operatives in developing countries exporting directly to the European market is increasing. Currently, 20% of 'fair trade' products in Italy are organic, certified by control bodies that are recognised by bodies of European countries. In Austria fair trade products showed growth rates between 24% to 27% in 2007 and 2008, showing that this niche still has potential (Holley-Spiess and Möchel, 2009). One of the biggest issues for such a niche market is the high certification costs. At the same time, one of the major risks for the competitiveness of national organic production is represented by imports from emerging countries.

In Italy there are about 1000 specialised shops (1/60,000 inhabit.), independent or franchises related to distribution channels for organic food. The most important franchise is NaturaSì. The specialised shops are mainly spread throughout Northern Italy (95). In 2004 the total value of sales in Italy was about 282 million euros.

In Italy the specialised shops such as national and regional franchising chains (NaturaSì, CarneSì, Verona, Italy; Bottega and Natura, Turin, Italy) still play a dominant role with a market share of 55% (Figure 10). Moreover, in the most important Italian supermarkets, organic products do not exceed 300 SKUs, while in specialised shops such categories rise up to 500.

In Germany the organic market has been mainly built up and dominated by specialised shops, but in recent years conventional supermarkets have entered the market and now have a market share of 49% (Figure 10). The growth rates for several distribution channels in Germany underline the importance of the retail sector and especially of the discount chains (Figure 11). While specialised shops grew by 7% in 2006, retail chains grew by around 34%, but from this 34% of growth around 74% belonged to the discount chains (and 17% to hypermarkets and 20% to small supermarkets; GfK Germany, 2007).

In contrast to Germany the retail chain REWE pioneered the Austrian organic market, when they introduced their organic private label 'Ja natürlich' back in 1994. The major proportion (about 66%) of Austrian organic products is distributed via conventional food retail chains. About 15%

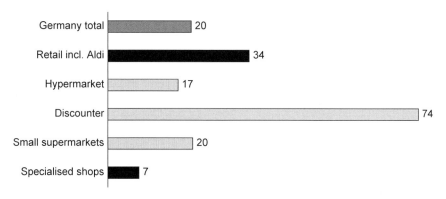

Figure 11. Growth rates (in percentages) of distribution channels for organic food in Germany (year 2006 to 2005) (GfK Germany, 2007).

is sold through specialised food shops. About 7% is distributed via direct marketing from farms. 7% are exported and around 5% are sold through home consumption (gastronomy, schools, canteens, hospitals). Farmers' markets can be found in every Austrian region (Holley-Spiess and Möchel, 2009).

In France, the association BioCoop organises the distribution of organic food through 170 selling points. In the last decade, supermarkets have gained more and more importance with respect to small natural food shops. Convenience stores (in French *Superettes*) are small specialised (bio-) supermarkets with self service with a surface of 200 to 500 square meter, they in part replace but also supplement the small specialised shops. Today, nearly 40% of the organic food is sold through supermarket chains, whereas the rest is sold through health food stores, direct sales, and open-air organic farmers markets (Figure 10). Most of the supermarket chains have their own organic food label, carrying an array of products from dry foods to dairy products, to meat and fresh products.

In 2004, the retail market for organic products in the United Kingdom was worth an estimated 1.78 billion euro, demonstrating a steady growth across the sector. Sales of organic products through direct and alternative markets, such as box schemes and independent retail shops, increased considerably during the year. Retail sales of organic products through the multiple retailers continued to grow, but at a much slower rate than in previous years. Nevertheless retailers have a market share of 75% in United Kingdom (Figure 10).

Next we examine some out-of-home channels (food service channels and organic canteens) distributing organic foods. In particular, the sector operators identify both an increasing penetration of organic products in the food service channel and out of home consumption as global trends for the coming years.

The Bio Bank 2004 database has detected 156 organic restaurants and 22 organic food service companies in Italy. In 2004, Ismea (2005) has monitored the Italian organic food service channel, carrying out research on a sample of 21 restaurants and 6 catering companies. The main differences between the two channels have been noticed in the organic food suppliers: for the restaurants the main suppliers are farms (59,8%), for the catering companies the main suppliers are specialised wholesalers (36,7%).

The additional cost of an organic meal compared to a conventional one corresponds to +32.5% for catering societies and +20.8% for restaurants.

From 1999 to 2004 in Italian canteens the use of flavouring and organic foods increased from 110 to 608 (+400%). The regional classification pointed out Emilia-Romagna as the leader with 119 'organic canteens', followed by Lombardia (105), Toscana (77), Veneto (69) and Friuli-Venezia Giulia (65). Moreover, starting from 2005, the Emilia-Romagna Regional Government and the Organic Producer's Association of Emilia-Romagna created an Internet website dedicated to organic foods and collective catering services: http://www.sportellomensebio.it/.

In the United Kingdom, government procurement policy and the 2004 Soil Association report on school meals have driven a strong emphasis for schools, hospitals and other public sector catering outlets to serve organic food. Over 300 schools throughout the UK have been in contact with the Soil Association about improving school meals.

In Austria, canteens and catering service kitchens in the public and private sectors are increasingly using organic food. Depending on the different provinces they are supplied by a high percentage of organic products. In Vienna, kitchens of all public institutions offer certain components in organic quality (up to 30%), and in Salzburg 50% of the food offered in kindergartens is organic. By now, the organic share of turnover of the Austrian food service operators in the public and private sectors, e.g. restaurants, hospitals, nursing homes, schools and canteens, sums up to around 44 million euro/year (Holley-Spiess and Möchel, 2009).

Even in the European countries where the conventional channel dominates, there is recently a growth of 'alternative' channels, such as organic catering (school canteens and hospital canteens), and short channel (direct sales, flea markets, box scheme) which combine the needs of regular consumers of organic products with those of small independent farmers (Berardini *et al.*, 2006).

5.5 Communication

In general the challenges for successful communication strategies are the same for organic products as for conventional food products. Due to the fact that the market is very heterogeneous and that there are no dominant global/international organic brands owned by the food industry, communication activities are mainly:
- low budget activities (below the line activities for e.g. events, fairs, promotions or direct marketing) from farmers associations or NGOs;
- umbrella marketing activities of governmental institutions or NGO's to promote organic agriculture/products in general;
- or branding activities of retail chains for their private labels.

For successful communication two questions have to be solved before starting a marketing campaign. First, what is the knowledge, what are the beliefs of the consumer about the product? Second, which communication media do they mainly use?

A study about Austrian consumers for example revealed that since more then 15 years, despite the fact that organic products are available in supermarkets, there are still major information deficits (AMA, 2007). Knowledge about criteria for organic agriculture or about control institutions, organic labels or organic brands is low. Claims such as 'environmental friendly agriculture' or 'directly from the farm' are wrongly associated with organic. Also the concept of integrated pest/crop management is often confused with organic agriculture. It is quite probable that consumers of other European countries have similar information deficits. Public relation campaigns from governmental institutions or NGOs should be one useful measure to reduce these information deficits.

Concerning the primary sources of information word-of-mouth and mass media are important sources (AMA, 2007; ISOE, 2003). Interestingly it seems that light users prefer other sources of information than heavy users of organic food. Light users mentioned information from media/advertising about healthy nutrition, organic farming etc. as the most important trigger for consumption of organic food. Heavy users mentioned word-of-mouth communication from friends or relatives in the first place (AMA; 2007). A reason for this difference could be that heavy users probably have higher product involvement and seem to be more critical, which results in a more active information search behaviour for trustworthy information sources, thus leading towards a stronger word-of-mouth orientation.

A study about German consumers reported that TV- and lifestyle magazines are the first source of information, followed second by word-of-mouth communication, and TV cooking shows on the third, women magazines and magazines with independent product tests on the following ranks (ISOE, 2003).

An example for umbrella marketing activities is for e.g. the communication campaign of the Italian Ministry of Agriculture (MIPAF):
- 'Organic agriculture grows the nature': a leaflet distributed in primary and secondary schools dedicated to organic agriculture.
- 'To be informed gives pleasure': 6 publications distributed during promotional exhibitions, also available on the MIPAF web site (www.politicheagricole.it). The publication dedicated to organic food is named 'Organic foods: quality and health at lunch'.

In Italy during the last years, the so-called bio-alliances have been created; with different agricultural and food companies developing a private label together. Examples of such 'organic alliances' are the Consortium Almaverde Bio (specialising in fruit and vegetables), the former alliance Verybio selling packed products (meat, fruit and vegetables), and Bioitalia created by the union of farms from Campania (packed products, wine, beverage, flavouring). The AlmaverdeBio Consortium of organic producers is strongly promoting an 'umbrella' brand through national TV commercials, available on the website http://www.almaverdebio.it.

Other alliances have emerged to sell organic products under an umbrella brand, such as the organic alliance between Alce Nero (cooperative of organic food producers) and Conapi (one of the biggest producers of honey) who have founded the company Mediterrabio, now Alce Nero & Mielizia.

In Western Europe, on the governmental level Germany invests substantial economic resources into the promotion of organic foods. The German Ministry of Agriculture dedicates 3 websites to organic food.

In Austria there is one umbrella organisation called Bio Austria (it opened its office in Vienna in 2003). For farmers and processors it offers services concerning advisory, quality and product management, research and innovation, and marketing. For consumers it offers specific information services. Around 10,000 consumers contact the organisation via email or telephone to get information about organic agriculture. A database with more than 50,000 consumers has been built up based on these customer contacts. Most consumer requests are about local shopping guides, direct selling organic farms, organic home delivery services or specialised organic stores. The importance of branding over mass media illustrates the success of the 'Yes naturally' advertising campaign. Since 2005 a speaking piglet and a farmer living on an organic farm are used as storytellers. For e.g. an advertising campaign for 'Yes naturally' fruit yoghurts brought a profit in turnover and market share of around 13% (Poschacher, 2009).

6. Conclusions

Organic agriculture represents a real opportunity for European farmers and it actively contributes to the vitality of rural areas. In many member countries of the EU, an increase of the organic sector gives rise to new possibilities of employment in agriculture, as well as in the processing industry and in related services. Thus, it is important for the economic development as well as for the social cohesion in rural areas.

The development stage of the national organic market in Europe is quite diverse: 'In countries like Germany, France, Austria, Belgium, Switzerland, Denmark, Italy, Luxemburg, the Netherlands,

United Kingdom and Sweden the organic market has already reached a high level of maturity (Van Osch *et al.*, 2008: 381). These mature markets are characterized by:
- High level of organic standards and market regulations.
- Well-developed organic market supply chains.
- Sustainable long-term growth of the organic food & drink market.
- Highly diversified products with excellent product availability in the main distribution channels.
- Strong organic private labels with a large product assortment.
- High customer reach and strong market penetration.
- Dedicated and focused marketing strategies of the main actors.
- Entry of new players such as specialised organic supermarkets offering only organic products
- Governmental support on the agricultural as well as market level (Van Osch *et al.*, 2008: 382).

In comparison to the before mentioned countries, less developed organic markets can be found in Eastern Europe but also in Portugal, Spain, Norway and Finland. These countries still have to develop efficient supply structures for the domestic organic food production. Mainly specialised organic stores are the market pioneers in these countries, followed by international retail chains, which offer a product assortment mainly based on imported products. Therefore the price levels of organic products are still quite high, resulting in a lack of consumer acceptance (Van Osch *et al.*, 2008: 382). Consumers in Austria, Switzerland and in Scandinavian countries are the biggest purchasers of organic food. These countries have over 5% of total food sales as organic food sales, which is the biggest market share of total food sales in Europe. In contrast Central & Eastern European consumers spend the least on organic food, with a market share below 1% (Sahota, 2009).

Distribution is a key factor for the growth of the organic sector. National and international retail chains play a significant role in establishing and developing organic markets all over Europe. The retail sector is not only the most powerful actor in the supply chain but also a strong innovator, due to its dominant role of its organic private labels. Specialised organic stores face tremendous competition, not only from conventional retail but also from new players such as organic supermarkets. For example in Germany specialised organic stores dominated the market for decades, but now the biggest share of sales is through supermarkets. Organic food sales from discount supermarket chains are especially growing significantly.

All organic markets strive to achieve economies of scale, which recently lead to a discussion about 'conventionalisation of organic farming'. With growing sales and sustained governmental support new farmers enter organic agriculture, farmers that are less 'idealistic' and more market oriented. Bigger farm sizes and a 'weakening' of organic standards are points of criticism. In 2008 the growing consumer demand also lead to supply bottlenecks, especially for products from domestic production, forcing wholesalers and retailers to import from third-countries outside of Europe. Food safety, transparency and long transport routes are issues gaining importance within a maturing organic market. Transporting organic strawberries in winter from Spain over 2,000 km to other European countries, poses the question how far this phenomenon is in accordance with the organic philosophy? As long as price margins show significant differences, food scandals due to producers/traders selling conventional food as organic food are a constant threat. The entrance of discount retail chains onto the organic market is a further sign for a more differentiated organic market in the future, splitting into a 'mass organic market' and a 'premium organic market'. Involving a lower price premium compared to conventional food – in many cases imported from third countries such as Turkey, or from Asia or Africa – and on the other hand premium organic food combined with other concepts such as the regional/local food concept, or with fair trade or with corporate social responsibility could be likely scenarios for a future European organic market. In the long and middle run – despite the slowing growth rates in 2008 due to the financial crisis – organic food and beverage will be an important growth factor for the food supply chain. A widening gap in income distribution of European consumers will further promote the emergence of a discount and premium organic market. The ageing society will continue to demand healthy food, as market studies show this is one of the core strengths of organic food in the mind of the consumers. A continuous climate change discussion in the media will further support environmental friendly production systems such as organic farming. Therefore it is of utmost importance to fulfil consumer needs for trustful, safe and reliable organic food production and marketing systems.

Discussion in lecture room

Apply your theoretical knowledge with respect to this paper:
- What does 'conventionalisation' of organic farming mean? Use Internet resources and scientific literature to summarise the debate on conventionalisation. Do you know any local/ national examples of conventionalisation?
- The major part of the worldwide demand for organic products is located in Europe and North America, while the three biggest production areas are in Australia, Europe and Latin America. Are long transport routes, resulting in a higher carbon footprint of organic products in accordance with the organic philosophy or in contradiction?
- Discuss the price formation along the organic food supply chain. Use Section 5.3 in this paper to check if the margins on production, wholesale and retail level are different compared to the price margins in conventional food supply chain.
- Formulate a marketing plan for the launch of a national organic brand. Read Section 4 and define a clear segmenting, targeting and positioning strategy. Make a store check in local supermarkets of existing organic brands and position your brand, in a new and differentiated way from the existing ones.

Acknowledgments

This paper was developed jointly by the authors. Nevertheless, the individual contribution may be identified as follows: Rainer Haas, Sections 2-5, Maurizio Canavari, Sections 5.5 and 6, Siegfried Pöchtrager, Section 5.1. Roberta Centonze and Gianluca Nigro collaborated to a previous version of this paper.

References

AMA (AgrarMarkt Austria), 2007. Hintergründe zur Meinungsbildung über biologische Lebensmittel. Available at: http://www.biologisch.at/images/stories/Leben/amabiostudie07.pdf. Accessed February 2008.

AMA (AgrarMarkt Austria), 2009. RollAMA, 3rd Trimester 2008. Bio Bank, 2004. Il Biologico in cifre. Forlì: Distilleria EcoEditoria. Available at: http://www.keyquest.at/produkte-services/lebensmittel/haushaltspanel-rollama.html. Accessed August 2009.

Anon, 2008. Bio fasst Fuss im traditionellen Lebensmitteleinzelhandel. Ergebnisse des LZ Trendbarometers Bio 2008. In: Deutscher Fachverlag (ed.), LZ Dossier Bio, Frankfurt am Main, Germany, pp. 52-57.

Berardini, L., Cianavei, F., Marino, D. and Spagnolo F., 2006. Lo scenario dell'agricoltura biologica in Italia. Working Paper SABIO no. 1, INEA – Istituto Nazionale di Economia Agraria, Rome, Italy.

Canavari, M., 2007. Current Issues in Organic Food: Italy. In: Canavari, M. and Olson, K.D (eds.), Organic Food. Consumers' Choices and Farmers' Opportunities. Springer, New York, Ny, USA, pp. 171-183.

Carrefour, 2009. Carrefour Discount. Our response to consumer expectations. Available at: http://www.carrefour. com/docroot/groupe/C4com/Pieces_jointes/Communiques_de_presse/2009/DP$Carrefour$Discount$- $fiches$-$version$VE.pdf. Accessed August 2009.

EC, 2005. Organic Farming in the European Union - Facts and Figures. General Direction Agriculture and Rural Development. European Commission, Brussels, Belgium. Available at: http://ec.europa.eu/agriculture/ organic/files/eu-policy/data-statistics/facts_en.pdf. Accessed January 2010.

FAO (Food and Agriculture Organization of the United Nations), 1999. Codex Alimentarius Commission Guidelines for the Production, Processing, Labelling and Marketing of organically produced foods. cac/gl 32-1999.

Fricke, A., 1996. Das Käuferverhalten bei Öko-Produkten – Eine Längsschnittanalyse unter besonderer Berücksichtigung des Kohortenkonzepts. Europäische Hochschulschriften: Reihe 5, Volks- und Betriebswirtschaft Bd. 1960, Peter Lang GmbH, Frankfurt am Main, Germany.

GfK Germany, 2007. Bio-Käufer in den Sinus Milieus – Eine Studie der GfK Panel Services im Auftrag des Bio Verlags. Germany. Available at: http://www.biohandel-online.de/public/2007/SinusGfKBioFach2007.pdf. Accessed March 2007.

Haas, R., Ameseder, C. and Liu, R., 2010. Factors influencing purchasing decisions of Austrian distribution channel operators towards 'made in China' organic foods. In: R. Haas, M., Canavari, B. Slee, C. Tong and B. Anurugsa (eds.), Looking east looking west: organic and quality food marketing in Asia and Europe. Wageningen Academic Publishers, Wageningen, the Netherlands, pp. 115-126.

Holley-Spiess, E. and Möchel, K., 2009. Handel macht mit Qualitätslabel weiter gute Geschäfte. Wirtschaftsblatt, Dienstag 21. April 2009, 6.

ISMEA/AcNielsen, 2005. L'evoluzione del mercato delle produzioni biologiche: l'andamento dell'offerta, le problematiche della filiera e le dinamiche della domanda. ISMEA, Rome, Italy.

ISOE, 2003. Zielgruppen für den Bio-Lebensmittelmarkt. Eine empirische Untersuchung des Instituts für sozial-ökologische Forschung im Auftrag der Geschäftsstelle Bundesprogramm Ökologischer Landbau. Available at: http://www.orgprints.org/4554. Accessed December 2007.

Kreuzer, K., 2009. Fachhandel in Österreich tritt auf der Stelle. Available at: http://www.bio-markt.info/web/ Europa/Oesterreich/Fachhandel_Oesterreich/84/98/0/5451.html. Accessed August 2009.

Lunati, F., 2006. Dallo sviluppo rurale una spinta alla crescita delle produzioni 'BIO'. Agricoltura, mensile dell'Assessorato Agricoltura Regione Emilia-Romagna 6: 48-50.

Nielsen, 2006. Trend Navigator Bio. Available at: http://www.acnielsen.de. Accessed January 2007.

Nielsen, 2007. Bio Studie. Eine Nielsen Studie mit Analysen aus Nielsen Handelspanel, Oktober 2007. Available at: at.nielsen.com/news/documents/BioCharts_Presse.pdf. Accessed August 2009.

Poschacher, R., 2009. Verbal information of the brand manager of Yes naturally. 31 August 2009.

Ratanawaraha, C., Ellis, W., Panyakul, V. and Rauschelbach, B., 2007. Organic Agri-business: A Status Quo Report for Thailand 2007. Bangkok, Thailand.

Sahota, A., 2009. The Global Market for Organic Food & Drink. In: H. Willer and L. Kilcher (eds.), The World of Organic Agriculture. Statistics and Emerging Trends 2009. FIBL-IFOAM Report. IFOAM, Bonn; FiBL, Frick; ITC, Geneva, Switzerland, pp. 59-63.

Schmid, O., Sanders, J. and Midmore, P., 2004. Organic Marketing Initiatives and Rural Development. OMIaRD Publication, Volume 7. University of Wales, Aberystwyth, UK.

Spiller, A.; Lüth, M. and Enneking, U., 2004. Analyse des Kaufverhaltens von Selten- und Gelegenheitskäufern und ihrer Bestimmungsgründe für/gegen den Kauf von Öko-Produkten. Selbstverlag – Bundesprogramm Ökologischer Landbau, Bonn, Germany.

USDA (United States Department of Agriculture), 2007. Organic Agriculture: 2007. The Census of Agriculture. Volume 1, U.S. National Level Data. Available at: http://www.agcensus.usda.gov/Publications/2007/Full_ Report/Volume_1,_Chapter_1_US/. Accessed January 2007.

Van Osch, S., Schaer, B., Strauch, C. and Bauer, C. (eds.), 2008. Specialised Organic Retail Report 2008. Practical Compendium of the Organic Market in 27 European Countries. Ora, Ecozept and bioVista, Vienna, Austria.

Willer, H. and Kilcher L. (eds.), 2009. The World of Organic Agriculture. Statistics and Emerging Trends 2009. FIBL-IFOAM Report. IFOAM, Bonn, Germany; FiBL, Frick; ITC, Geneva, Switzerland.

Willer, H., Rohwedder, M. and Wynen, E., 2009. Organic Agriculture Worldwide: Current Statistics. In: H. Willer and L. Kilcher (eds.), The World of Organic Agriculture. Statistics and Emerging Trends 2009. FIBL-IFOAM Report. IFOAM, Bonn, Germany; FiBL, Frick; ITC, Geneva, Switzerland, pp. 25-58.

Zanoli, R., Bähr, M., Botschen, M., Laberenz, H., Naspetti, S. and Thelen, E., 2004. The European consumer and organic food. - Organic marketing initiatives and rural development Vol. 4, University of Wales, Aberystwyth, UK.

Small farms and the transformation of food systems: an overview[1]

E.B. McCullough[1], P.L. Pingali[1] and K.G. Stamoulis[2]
[1]Bill & Melinda Gates Foundation, P.O. Box 23350, Seattle, WA 98102, USA
[2]FAO, Agricultural Development Economics Division, Viale delle Terme di Caracalla, 00153 Rome, Italy

Abstract

This article illustrates the impact of globalization on food supply chains in developing countries and its specific impact on smallholder farmers. Global shifts in consumption, marketing, production and trade are leading to organizational changes along the food chain. The authors present a framework for evaluating impacts at the household level and describe three different typologies for food systems that correspond with the development process. Policy makers face the challenge to manage the transition for smallholder households, which focuses on linking smallholders into modern food chains, upgrading traditional markets and providing exit strategies for those who are marginalized by the transformation process.

1. Introduction

By making a strong case for the importance of agriculture in poverty reduction even in developing countries with largely urbanized populations, the 2008 World Development Report has continued the renewed interest in agriculture as a force for poverty reduction (World Bank, 2008). Research has shown that rural poverty reduction, resulting from better conditions in rural areas and not from the movement of rural poor into urban areas, has been the engine of overall poverty reduction (Ravallion *et al.*, 2007). Organizational changes that are currently underway in developing-country food systems necessitate a new look at agriculture's role in poverty reduction with an eye on the changing rural economy. The reorganization of supply chains, from farm to plate, is fuelling the transformation of entire food systems in developing countries. With the changing rural context in mind, we revisit prospects for poverty reduction in rural areas, particularly in the small farm sector. The transformation of food systems threatens business as usual but offers new opportunities for smallholder farmers and the rural poor.

The purpose of this paper is to take stock of important trends in the organization of food systems and to assess, with concrete examples and case studies, their impacts on smallholder producers in a wide range of contexts. This paper brings together relevant literature in a consistent manner and examines more holistically the issue of changing food systems, moving beyond the focus of supermarkets, which has been a dominant concern in recent literature. We focus on domestic markets as well as exports, and on a wide range of sub-sectors, not just fresh fruits and vegetables and dairy. This paper begins with a description of changing consumption patterns in developing countries. Then we highlight organizational changes that have taken place along the food chain, recognizing important differences between countries, and exploring interactions between traditional and modern chains in countries where food systems are transforming. We present a framework for evaluating impacts at the household level, pulling together empirical evidence in support of the framework. We close with a policy discussion on managing the transition for smallholder households, which focuses on linking smallholders into modern food chains,

---

[1] Reprinted from Chapter 1 in the book *The transformation of agrifood systems: globalization, supply chains and smallholder farmers*, edited by E.B. McCullough, P.L. Pingali and K.G. Stamoulis, with permission from Food and Agriculture Organization (FAO) of the United Nations.

upgrading traditional markets and providing exit strategies for those who are marginalized by the transformation process.

2. The transformation: an overview

In this paper, we lay out three different typologies for food systems that correspond roughly with the development process. The first is a traditional food system, characterized by a dominance of traditional, unorganized supply chains and limited market infrastructure. The second is a structured food system, still characterized by traditional actors but with more rules and regulations applied to marketplaces and more market infrastructure. In structured food systems, organized chains begin to capture a growing share of the market, but traditional chains are still common. The third type is an industrialized food system, as observed throughout the developed world, with strong perceptions of safety, a high degree of coordination, a large and consolidated processing sector and organized retailers.

Major global shifts in consumption, marketing, production and trade are brought about, above all, by four important driving forces associated with economic development: rising incomes, demographic shifts, technology for managing food chains and globalization. As these changes are played out, modern chains capture a growing share of the market, and food systems transform. The variable that differs most strikingly between food system typologies is the share of the food market that passes through organized value chains. We identify economic factors that explain how modern chains capture a growing share of food retail over time, and we explore specific differences between organized and traditional chains. Then we examine the implication of the spread of modern chains from the perspective of chain participants and with respect to the entire food system. In practice, the boundaries between these food system typologies are not easily discernible. Nor is the path from traditional to structured to modern a linear one. A mix of different types of chains can be found within one country depending on the commodity involved, the size of urban centres and linkages with international markets (Chen and Stamoulis, 2008).

Understanding how different types of chains relate to each other is important for predicting future opportunities for smallholder farmers as food systems reorganize. In developing countries, the food system is typically composed of domestic traditional chains, domestic modern chains and export chains, which are usually exclusively modern. When traditional marketing systems fail to meet the needs of domestic consumers and processors, modern retailers develop mechanisms for bypassing the traditional market altogether. Modern food chains in developing countries advance rapidly due to global exposure, competition and investment, while traditional chains risk stagnation due to underinvestment. As the gap between traditional and modern food chains grows ever wider, the challenge of upgrading traditional chains becomes more pronounced. The entire food system's transition from traditional to structured is hindered as resources and attention are diverted from upgrading traditional markets in favour of bypassing them.

Assessing the full implications of changes for rural communities and, in particular, smallholder agriculture, requires an analysis of how risks and rewards are distributed both in traditional food systems and modern ones. As production and marketing change, there are obvious implications for smallholder farmers via changes in production costs, output prices and marketing costs. But changes in processing, transport, input distribution and food retail also impact rural households via household incomes (e.g. labour markets, small enterprises) and expenditures (e.g. food prices).

From farm to plate, one overarching trend is the rising need for coordination in modern food systems relative to traditional ones, and the transaction costs that are introduced as a result.

Coordination helps to ensure that information about a product's provenance travels downstream with the product. It also helps to ensure that information about consumer demand and stock shortages/surpluses is transferred upstream more efficiently to producers (King and Phumpiu, 1996). Improving coordination along the supply chain reduces many costs but introduces new ones (Pingali *et al.*, 2007). We explore and evaluate different strategies for coordination later in this paper.

2.1 Towards dietary diversification

Brought about by rising incomes, demographic shifts and globalization, dietary change is sweeping the developing world. Consumers are shifting to more diverse diets that are higher in fresh produce and animal products and contain more processed foods. Shifts in food consumption parallel income growth, above all, which is associated with higher value food items displacing staples (Bennett's Law). The effect of per capita income growth on food consumption is most profound for poorer consumers who spend a large portion of their budget on food items (Engel's Law). A sustained decline in real food prices over the last 40 years has reinforced the effect of rising incomes on diet diversification.

Per capita, incomes have risen substantially in many parts of the developing world over the past few decades. In developing countries, per capita income growth averaged around 1% per year in the 1980s and 1990s but jumped to 3.7% between 2001 and 2005 (World Bank, 2006). Growth rates have been most impressive in east Asia and slightly less spectacular in south Asia. Declining growth rates have been reversed since the 1990s in Latin America and since 2000 in sub-Saharan Africa. Income growth has been accompanied by an increase in the number of middle class consumers in developing countries, particularly in Asia and Latin America, whose consumption patterns have diversified (Beng-Huat, 2000; Solamino, 2006).

Beyond income growth, dietary diversification is also fuelled by urbanization and its associated characteristics, rising female employment and increased exposure to different types of foods. Globally, urban dwellers outnumbered rural populations during 2007 (Population Division of UN, 2006). Feeding cities is now a major challenge facing food systems. Female employment has at least kept pace with population growth in developing countries since 1980 (World Bank, 2006). Female employment rates have risen substantially in Latin America, east Asia, and the Middle East and north Africa since the 1980s.

Urban consumers typically have higher wage rates and are willing to pay for more convenience, which frees up time for income-earning activities or leisure. Therefore, they place a higher premium on processed and pre-prepared convenience foods than do rural consumers (Popkin, 1999; Regmi and Dyck, 2001). Rising female employment also contributes to this phenomenon (Kennedy and Reardon, 1994). Smaller families are typical of urban areas, so households can afford more convenience in terms of processed and prepared foods.

Globalization has led to increased exchanges of ideas and culture across boundaries through communication and travel, leading to a tightening of the global community which is reflected in dietary patterns, such as increased consumption of American style convenience foods. Urban consumers are exposed to more advertisements and are influenced by the wide variety of food choices available to them (Reardon *et al.*, 2008).

Dietary changes have played out differently in different regions and countries, depending on their per capita incomes, the degree of urbanization and cultural factors. The most striking feature of dietary change is the substitution of traditional staples for other staple grains (i.e. rice for wheat in east and southeast Asia) and for fruits and vegetables, meat and dairy, fats and oils (Pingali, 2007). Per capita meat consumption in developing countries tripled between 1970 and 2002, while milk consumption increased by 50% (Steinfeld and Chilonda, 2005). Dietary changes are most striking in Asia, where diets are shifting away from rice and increasingly towards livestock products, fruits and vegetables, sugar and oils (Pingali, 2007). Diets in Latin America have not changed as drastically, although meat consumption has risen in recent years. In sub-Saharan Africa, perhaps the biggest change has been a rise in sugar consumption during the 1960s and 1970s (http://faostat.fao.org/). Cereals, roots and tubers still comprise the vast majority of sub-Saharan African diets, and this is expected to continue into the foreseeable future (FAO, 2006). Total food consumption in developing countries is projected to increase in coming decades, so dietary diversification does not necessarily imply that per capita consumption of any food products will decline in absolute terms (Figure 1). However, by 2030, absolute decreases are expected in per capita consumption of roots and tubers in sub-Saharan Africa and of cereals in east Asia (FAO, 2004). Since cereals are used as inputs in animal production, total cereal demand will not decrease due to indirect consumption.

2.2 Trends in food systems organization

Consumption of higher value products is on the rise in developing countries, and supply chains are ready to meet these demands. But which chains will reach dynamic consumer segments in developing countries, and which farmers will supply these chains? From farms to retail, technology and 'globalization' are the most important drivers of reorganization of the chains linking producers and consumers. Innovations in information and communications technology have allowed supply chains to become more responsive to consumers, while innovations in processing and transport have made products more suitable for global distribution. Technological innovation in food supply chain management has arisen in response to volatility in consumer demand (Kumar, 2001). New communication tools, such as the Universal Product Code, which came on line in the 1970s, have improved the efficiency of coordination between actors along

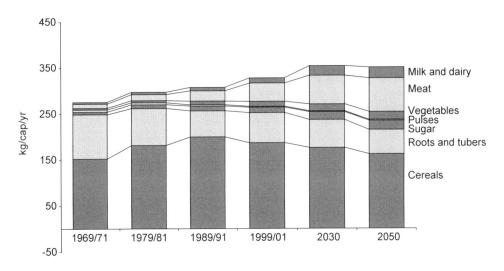

Figure 1. Trends and projections for dietary diversification in east Asia (FAO, 2006).

the supply chain to shorten response times to demand fluctuations (King and Venturini, 2005). Packaging innovations throughout the second half of the 20th century continued to extend food products' shelf lives (Welch and Mitchell, 2000). Meanwhile, a downward trend in transportation costs and widespread availability of atmosphere-controlled storage infrastructure have made it cost effective to transport products over longer distances. Crop varieties have been tailored specifically to chain characteristics, for example to meet processing standards or to extend shelf life. Conventional breeding and, more recently, biotechnology, have allowed these shifts.

'Globalization' in retail and agribusiness is marked by liberalization of trade as well as of foreign direct investment (FDI). Trade has maintained a constant share in global food consumption but is shifting towards higher value products, such as processed goods, fresh produce and animal products (Hallam *et al.*, 2004). Flows in capital can impact food systems as profoundly as flows in products. Rising Foreign Direct Investment (FDI) flows into developing countries have been linked with concentration throughout the food industry, boosts in productivity and innovation, and an increase in non-traditional agricultural exports (Wilkinson, 2008). Foreign direct investment in agriculture and the food industry grew substantially in Latin America and in Asia between the mid-1980s and mid-1990s, although investment remained very low in sub-Saharan Africa (FAO, 2004). In Asia, FDI in the food industry nearly tripled, from 750 million to 2.1 billion US$ between 1988 and 1997. During that same period, food industry investment exploded in Latin America, from around 200 million to 3.3 billion US$. There is a limit in the availability of sector-specific FDI data since 2001, but economy-wide data through 2005 show a similar pattern: with long-term increases in developing countries in Asia and Latin America, with 2002-2003 slumps in both cases, and with Africa lagging behind but growing somewhat steadily since the 1970s (Figure 2). FDI flows into Africa have lagged behind those into Asia and Latin America because of structural and institutional constraints. The world's least developed countries (LDCs) receive only 2% of global foreign direct investment.

The transformation of food systems is not something that occurs overnight. While many of the factors that affect food systems can change rather quickly, their reorganization involves large

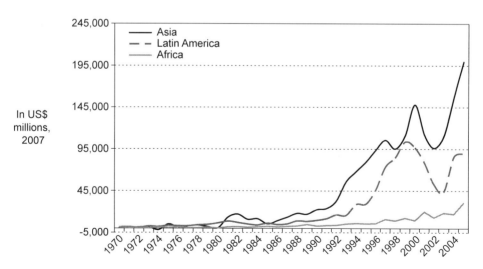

Figure 2. Annual Foreign Direct Investment (FDI) net inflows into developing countries by region from 1970 (OECD, 2007).

investments in specialized infrastructure, institutional change and regulatory reform. Often, these components are jointly determined rather than one causing the others. Institutions evolve as modern systems expand and infrastructure is built to accommodate the needs of the evolving markets and players. As mentioned above, multiple typologies can be observed simultaneously in the same country. Within one country, organizational change may take place earlier in chains for products that are prone to safety violations, such as meat and dairy (Chen and Stamoulis, 2008). International concerns over trans-boundary diseases (e.g. avian flu) place pressure even on non-exporting countries to upgrade supply chains in order to reduce the incidence of outbreaks and allow for better response when outbreaks occur. When a country does export food products, the onus is on the exporters to demonstrate that their products (and/or the production and post-harvest systems that give rise to them) meet the importers' safety standards.

Country typologies by stage of transformation

Organizational changes in food systems vary in speed and extent across contexts (national, sub-national, type of product and chain), and impacts vary across households and household typologies. At the country level, perhaps the most important determinant of the transition is the country's position in the agricultural development process. This is the path by which, over time, per capita incomes rise as the share of agriculture in a country's work force and economy decline (Figure 3) (Pingali, 1997, 2006).

Countries at the low end of the transformation process are characterized by low per capita incomes, with the agricultural sector accounting for more than 30% of the national Gross Domestic Product (GDP) and over 50% of the work force (http://faostat.fao.org/). In these countries, which are mostly located in sub-Saharan Africa and include, for example, Zambia, Kenya and Uganda,

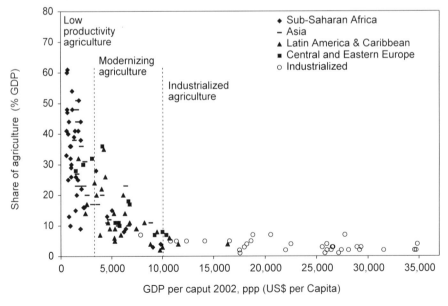

Figure 3. Falling share of agriculture in the economy as Gross Domestic Product (GDP) rises. (World Bank, 2006).

agriculture is mainly oriented towards the production of non-marketed staples, and cropping systems are often diversified at the farm level with inputs generated on the farm (Table 1). Some of the surplus production is marketed, but production systems are mainly subsistence-oriented. Staple crop productivity growth remains the primary engine of overall economic growth. In traditional agricultural economies, the transformation of food systems has been slow to take off. High value, organized retail establishments may cater to a limited, often expatriate, clientele in capital cities, but most supply chains for most crops are still traditional in nature. Developing modern, vertically integrated supply chains is difficult and expensive in countries with poor road infrastructure and failed institutions.

In Africa, there is a high degree of dualism between traditional domestic food chains and organized chains, whether domestic or export-oriented. Food safety standards for poor consumers, who frequent traditional markets, are quite low as those markets are largely informal. Another major problem is the vicious cycle of low surplus volumes constraining market development, which then reinforces the subsistence nature of low-input production systems. In sub-Saharan Africa, continued underinvestment of public goods supporting smallholder agriculture is likely to further widen the gap between traditional domestic markets and the formal processing and retail sector (Jayne, 2008). Transforming consumer demand, particularly in urban areas, will be met with imports for products that domestic supply chains cannot provide competitively. In sub-Saharan Africa, urban demand is increasingly met with imports rather than by domestic producers. According to urban consumer surveys in Mozambique, Kenya and South Africa, expenditures on wheat and/or rice were higher than those on maize (Jayne, 2008). Without proper linkages

Table 1. Characteristics of food systems by country typology.

	Traditional	Structured	Modern/integrated
Share of agriculture in GDP	High	Medium	Low
Urbanization	Rural	Urbanizing to varying degrees	Urbanized
State of the agricultural economy	Traditional	Modernizing	Industrialized
Rural income sources	Few opportunities outside of agriculture (farming or ag wage labour), high migration	More diversified opportunities, dualistic	Agriculture and manufacturing, dwindling rural population
Agriculture's role in poverty reduction	Agriculture growth stimulates mass poverty reduction via market linkages and labour for traditional export commodities	Agriculture growth reduces rural poverty and manages the urban transition. Opportunities in processing and high-value crops in domestic markets	Agriculture growth promotes rural income parity, agribusiness provides employment, provision of ecosystem services
Institutions	State boards	Transitioning	Regulatory
Examples	Bhutan, Kenya	India, China, Honduras, Mexico	US, EU

between rural producers and urban consumers, economic growth in urban areas cannot spur widespread rural poverty reduction.

In modernizing economies, the agricultural sector accounts for a 10-30% share of the economy and a 15-50% share in the work force. Modernizing economies, which are mostly located in Asia, Latin America and central and eastern Europe, vary greatly with respect to urbanization rates and income[2]. Countries in Asia, such as Vietnam and Bangladesh, typically have lower urbanization rates, while those in Latin America and central and eastern Europe, such as Mexico and Honduras, have higher urbanization rates. In modernizing economies, the majority of farmers produce for domestic markets; but both subsistence- and export-oriented systems are present. Food systems in modernizing economies are neither traditional nor industrialized but somewhere in between. The more urbanized economies of central and eastern Europe and Latin America will be marked by more opportunities for marketing high-value products domestically. High rural poverty rates underscore the importance of agricultural growth for improving rural incomes in many Asian countries with lower urbanization rates. Meeting urban food demands can be the new source of growth for these economies. Further improving diversification into higher value agriculture to meet domestic urban demand is an important goal.

In industrialized economies, such as the US, the EU, Australia and New Zealand, agriculture usually accounts for less than 10% of GDP and less than 15% of the work force. Markets are domestically and internationally oriented; output mixes are highly diversified with a well-developed processing sector providing opportunities for value addition. Typically, industrialized agricultural systems are highly mechanized and scale economies are quite pronounced. Differentiated products flow through well organized value chains, and commodity markets maintain basic safety standards through regulation (Kinsey and Senauer, 1996).

Apart from the phase in which a country finds itself in the agricultural development process, several other factors can influence the speed and nature of the transformation of food systems (although such factors usually correlate highly with the transformation process and a country's attractiveness to outside investors). It is important to remember that capital is mobile and policies at the national level are important determinants of the investment climate, which is affected by institutions, infrastructure, capacity and transaction costs (Bénassy-Quéré *et al.*, 2007; Globerman and Shapiro, 2002). Stable governments and institutions provide a better environment for large capital investments; widespread graft and excessive bureaucracy can discourage investment. Agribusiness firms looking to vertically integrate their supply chains will prefer countries where the regulatory environment is transparent and easy to negotiate. They will seek places where arbitration costs are low and coordination is easy to manage. All of these factors, which could be considered transaction costs, influence the cost of developing and managing supply chains, and therefore the competitiveness of their final products.

---

[2] According to the World Bank's classifications in the World Development Report, modernizing economies fall into two categories: 'transforming' and 'urbanized' (World Bank, 2008). Relative to the urbanized economies, transforming ones are marked by a greater share of agriculture in the work force, a lower GDP per capita, lower urbanization rates, higher overall poverty and rural poverty rates. It is not necessarily implied, though, that agricultural economies must pass from agricultural to transforming to urbanized rather than directly from agricultural to urbanized.

2.3 Organizational trends along the value chain

Acting at once and often reinforcing each other, driving forces have exacted and continue to exact major changes on food distribution systems. A wide body of literature, particularly from the last decade, describes the reorganization that has taken place in food chains, with implications for chain participants and for the broader economy (Table 2). Much of the evidence available is focused on retail, particularly supermarkets. Many of the procurement and marketing studies focus on fresh fruits and vegetables grown for export to consumers in developed countries. The dairy subsector has also garnered a fair amount of attention. While not all locations, crops or stages in the supply chain have received interest proportional to their importance for rural poverty, there nevertheless exists a robust set of documented studies from which to draw conclusions about the implications of the reorganization of supply chains and resulting transformation of food systems for food security, and in particular, rural poverty.

Retail consolidation trends

The proliferation of supermarkets in developing countries is one of the most widely cited elements of food system transformation. Trends in consumption pave the way for consolidation in the retail sector, which then reinforces dietary changes. Demand for safe food and for processed food

Table 2. Trends in the organization of food systems from farm to plate.

	Traditional	Structured	Industrialized
Consumption	Rising caloric intake, diversification of diets	Diet diversification, shift to processed foods	Higher value, processed foods
Retail	Small scale, wet markets	Spread of supermarkets, less penetration of FFV[1]	Widespread supermarkets
Processing	Limited processing sector	Processing offers employment and value addition opportunities	Large processing sector for domestic and export markets
Wholesale	Traditional wholesalers, with retailer bypassing for exports	Traditional and specialized wholesalers, some retailer bypassing	Specialized wholesalers and retailer bypassing through distribution centres
Procurement	Via traditional markets	Via structured (regulated) markets	Via managed chains, advance arrangements
Production systems	Diversified, low input systems	Intensive input use, specialization of cropping systems	More focus on conservation
Safety in food system	No traceability	Traceability in some chains with private standards	HACCP[2] system, private safety standards and public accountability (liability)
Vertical coordination	Relationships	Relationships/rules	Binding agreements, ICT[3] systems for efficient consumer response

[1] FFV = fresh fruit and vegetable.
[2] HACCP = hazard analysis and critical control point.
[3] ICT = information and communications technology.

products provides an entry point for organized, larger scale retail outlets in urban markets. By offering a wide variety of products, supermarkets can stimulate new demand through availability and exposure. Families who own refrigerators and vehicles are able to make fewer, but higher volume, trips to purchase food, which explains the strong link between the spread of supermarkets and the rise of the middle class. Income growth is closely linked with ownership of durable goods, like refrigerators and vehicles (Filmer and Pritchett, 1999).

The spread of supermarkets has been documented in a variety of studies specific to countries and regions (see Dries *et al.*, 2004; Hu *et al.*, 2004; Reardon and Berdegue, 2002; Weatherspoon and Reardon, 2003). Structural transformation of the retail sector took off in central Europe, South America and east Asia outside China in the early 1990s. The share of food retail sales by supermarkets grew from around 10% to 50-60% in these regions. By 2002, in central America and southeast Asia, the shares of food retail sales accounted for by supermarkets reached 30-50%. Starting in the late 1990s and early 2000s, substantial structural changes taking place in eastern Europe spurred growth in supermarkets, which now comprise 30-40% of food retail (Dries *et al.*, 2004). So far, supermarkets have failed to capture a large portion of food retail in south Asia (1-2%), China (11%), and Africa (with the exception of South Africa, 5-10%), despite the high growth rates that have been reported in the organized retail sector (Traill, 2006). There are indications of a rapid rise in supermarket growth rates in China and India over recent years.

A recent study by Traill (2006) involved compilation of a cross-country dataset on supermarket penetration in developing and developed countries (Figure 4). Using a multivariate regression, differences between countries were explained by per capita income, urbanization rates, female participation in the work force and income inequality. All of these factors were positively correlated with the share of food retail captured by supermarkets. It is important to stress that,

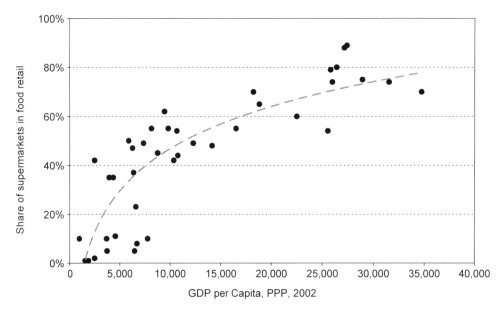

Figure 4. Rising Gross Domestic Product (GDP) per capita is associated with a larger share of supermarkets in food retail (Traill, 2006; World Bank, 2006).

in most developing countries with traditional food systems, supermarket share in retail is still limited to the 10% range, even lower for the fresh produce market segment. Low-penetration countries are unlikely to become high-penetration ones in the coming decade, even though supermarkets in developing countries have sustained impressive growth rates (Traill, 2006).

The methodology for collecting data on supermarket penetration differs from country to country, but supermarkets are usually defined as single, self-service retail outlets which exceed a threshold number of cash registers (e.g. 2-3) or floor space (e.g. 150m^2) (see the definition used in Neven and Reardon, 2008). Some important changes in the retail sector could go undetected in estimates based on such definitions, such as shifting procurement patterns among small-scale, traditional retailers of fresh and processed foods, which would have implications for the producers who supply them. Also, estimates of average retail share mask differences between sub-sectors. Supermarkets' share of fresh produce retail, for example, is consistently 25-50% of supermarkets' share of total food retail (Berdegue *et al.*, 2005).

Implications of consolidating retail

The implications of retail transformation on producers have been explored through a wide range of studies. Ultimately, small farmers are impacted through changing points and terms of sale and changing safety and quality requirements for products that are purchased. Two questions pertain: which farmers participate? What happens to those who cannot or do not? The clearest mechanism by which retail transformation impacts farmers is through changing procurement standards, particularly with respect to quality and safety of products (see Balsevich *et al.*, 2003; Berdegue *et al.*, 2005; Reardon *et al.*, 2008; Swinnen, 2007). In order to ensure year-round availability of produce, private retailers may specify delivery standards for minimum monthly shipments throughout the growing season. The delivery requirements can serve as a barrier to smallholders directly supplying retailers (Dolan and Humphrey, 2002). Other retailer standards relate to the products themselves or the methods by which they are produced, for example, the EurepGAP (Originally the European Retailer Produce Working Group Good Agricultural Practices) programme (McCluskey, 2007).

When possible, supermarkets have procured through regional distribution centres with the capacity to receive shipments from farmers, bulk-up orders, sort products and ship to retail locations. Since supermarkets' distribution centres perform wholesale functions, they are discussed in Section 9.

The practice of farmers selling directly to retailers is more common in fresh fruit and vegetable chains than in others. Evidence shows that supermarkets are more likely to procure directly from farmers or farmer groups in countries where supermarket penetration is still quite low, such as Thailand (Chen and Stamoulis, 2008). Relatively few farmers, small or large, supply supermarkets directly, so quality and safety standards are transmitted to farmers via processors, wholesalers and traders. The transmission of consumer preferences and retailer standards to producers depends on a number of factors, including the structure of the wholesale market.

Retail concentration in developing countries has implications for retail-related employment and for consumers. It is likely that at least some traditional retailers will be displaced by growth in the supermarket sector, leading to a net job loss in the retail sector. This hypothesis is based on the assumption that supermarkets are more capital intensive (with respect to labour) than are traditional retailers (Dries, 2005). Transformation of retail may cause consumer prices to go up or down, depending on the competitiveness of the sector, but the availability of more variety and more quality differentiation will improve consumers' welfare as long as prices are competitive. Consumers are likely to benefit from competition between organized and traditional retailers, as well as that within the supermarket sector, which can lead to improved services overall. In India, there is concern among the public that a change in FDI policy will drive small retailers out of business by offering low prices initially, at a loss. Predatory pricing patterns have been documented in many developed countries, with impacts borne by small retailers who must compete for customers and for suppliers (Foer, 2001; Reardon and Hopkins, 2006). Across Asia, there is evidence that consumers still prefer to buy their produce in traditional markets, where it tends to be fresher (see Chen *et al.*, 2005; Dirven and Faiguenbaum, 2008; Maruyama and Viet Trung, 2007; Singh, 2008). In Latin America, the small scale retail sector has relied on responsiveness to consumers to maintain some resilience in the face of competition with large retailers (D'Andrea *et al.*, 2006). Supermarkets may earn lower profit margins on fresh produce relative to other items, but offering it is important for improving loyalty amongst customers who place a premium on one-stop shopping.

The size of the urban middle class determines the nature of the retail clientele (Wilkinson, 2008). The more 'mainstream' domestic supermarket chains become, the more they must compete amongst themselves on price, product safety and quality, and with traditional wholesalers on price and freshness. Price and convenience have been common entry points for supermarkets in developed countries. The consumer base will determine customers' willingness to pay for quality, and how retailers should handle the trade-off between quality and price (Maruyama and Viet Trung, 2007). Modern retailers can out-compete traditional chains on food safety because they can implement traceability and communications technology. It appears, for now, that traditional chains can compete with organized ones on freshness and price. As the middle class grows, so will the number of organized retailers that cater to them, offering different combinations of quality, safety and economy based on consumer preferences. Asian consumers appear to be willing to tolerate a lack of traceability in modern chains, but a big public safety scare could boost demand for safe food. Absent a marketing opportunity posed by changing public perceptions or a regulatory shift, supermarkets will continue to procure through the traditional wholesale system.

More processing and trade in processed products

Processed products are capturing a growing value share in global agricultural production and trade at the expense of bulk commodities (Regmi and Dyck, 2001). Some higher income developing countries match or surpass global trends, but most least developed countries have not shared in opportunities to expand agricultural processing for domestic consumption or for export (FAO, 2006; World Bank, 2008). While the LDCs comprise 10% of the world's population, they account for only 0.4% of global manufacturing value addition in all sectors (Wilkinson, 2008). Yet in these LDCs, the food industry often accounts for the largest share of manufacturing value addition. In 17 of Africa's LDCs, over 80% of the manufacturing is in the agri-food sector. In most of Africa's other LDCs, the share is over 50%.

This points to the opportunity to expand food manufacturing for domestic and export markets using domestically grown raw materials. Furthermore, quality standards for raw materials to be used in manufacturing are often not as strict as those for fresh produce. Because of lower costs of compliance, and less seasonal price variability, scale economies have been shown to be less prohibitive, and so processing channels may be more accessible to smallholders than fresh produce channels. Quality standards for green bean canning firms are much lower than those for fresh green beans in Kenya, even though both products are destined for export. As a result, the green bean processing chain has sustained smallholder participation much better than the fresh green bean chain has (Narrod *et al.*, 2008).

With the transformation of food systems there has been a trend of upgrading and consolidation in agri-processing firms in developing countries. A shake-out of domestic processing firms has been observed with the entry of foreign firms, facilitated by the liberalization of FDI and trade (Chen and Stamoulis, 2008; Wilkinson, 2008). Competition to meet cost-effectively retailers' standards has led to consolidation amongst the remaining domestic processing firms in developing countries. Consolidation of retail in developing countries has been most pronounced in sub-sectors whose processors are smaller, more independent and less advanced technologically (Chen, 2004). Small processors in developing countries reported difficulties selling to supermarkets because the retailers applied large stocking fees (Chen and Stamoulis, 2008). For example, in southeast Asia, supermarkets have catalysed major changes in the fresh fruit and vegetable packing industry but not on chicken packers, who had already adopted internationally-accepted standards by the time domestic supermarkets became important buyers.

Bypassing of traditional wholesalers

By assembling large volumes of produce from a 'marketshed', wholesalers are better positioned to meet retailers' and processors' requirements than are individual farmers, particularly smallholders. In traditional chains, which are still widely prevalent in agriculture-based economies, farmers and traders supply traditional wholesalers, who then sell to individual retailers and processors, many of whom are small in scale. In modern chains, farmers and traders supply specialized wholesalers or distribution centres, who then sell to organized retailers and processors. In countries with modernizing food systems, both chains may exist side to side, with some exchange between them as conditions allow (Figure 5).

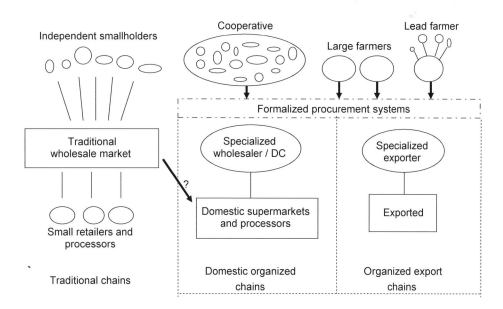

Figure 5. Interactions between traditional and organized chains in modernizing food systems.

When organized retailers first enter a country, they typically set up their own direct procurement systems. Specialized regional distribution centres are constructed to serve their wholesale needs when traditional and even specialized wholesale markets cannot, and once sufficient economies of scale are present. To justify the cost of a regional distribution centre for fresh produce with the savings generated, it is estimated that a retailer needs a minimum of 20 supermarkets (Chen and Stamoulis, 2008). In the developing world, there are few regional distribution centres for fresh produce outside of Latin America. In the US and Japan, major retail distribution centres began to displace wholesalers by the 1960s (Chen and Stamoulis, 2008). More recently, there is evidence in developed countries of a shift back towards direct procurement, at least in niche markets where consumers place a premium on fresh, local and seasonal produce.

Wholesalers assemble, grade and sort produce, bridging the scale gap between producers and retailers. Traditional wholesalers can differentiate products on the basis of basic functions, like size, colour and other easily observable characteristics. But product information that is not readily observable does not transmit well through the traditional system. Specialized wholesalers are better positioned to keep track of quality information and meet more exacting demands from retailers and processors (Golan *et al.*, 2004; Unnevehr and Roberts, 2002). For this reason, specialized wholesalers have captured important market segments in developing countries (Coe and Hess, 2005; Reardon and Berdegue, 2002; Reardon *et al.*, 2008). For organized retailers, bypassing traditional wholesalers affords better quality control and can lower costs if savings from reduced spoilage offset the costs of managing the distribution facility. High spoilage rates in traditional wholesale markets in developing countries give retailers strong incentives to bypass traditional wholesalers.

To date, evidence of bypassing has been limited in developing countries outside of Latin America (Chen and Stamoulis, 2008). With fresh fruits, vegetables and many bulk commodities, traditional wholesale markets remain vibrant, even in Latin American where supermarket penetration is high. In southeast Asia, specialized wholesalers have a small market share but play an important part in quality segmentation (Caldhilon *et al.*, 2006). The traditional market system can accommodate some quality differentiation, but it is inefficient if quality-differentiated prices are not transmitted to producers and quality information is not transmitted to consumers (Digal, 2004). There is evidence that traditional wholesalers are responding to the spread of specialized wholesalers and distribution centres by investing in upgrades to improve quality and safety. In Chile, public and private investments were made to improve traditional wholesale markets serving Santiago, with the goal of helping wholesalers compete with private chains on quality, safety, and customer service (Dirven and Faiguenbaum, 2008). Traditional markets may not be able to compete with specialized markets cost-effectively with regards to quality and safety.

From the perspective of the small-scale farmer, an important question is: who will the specialized wholesalers buy from, given the choice? And in order to compete with specialized wholesalers, will traditional wholesalers impose standards that lead to exclusion of smallholders? As retailers begin to pull a meaningful share of the wholesale market into distribution centres, what are the implications for smallholder farmers? The degree of duality between the traditional and specialized wholesale systems will determine, in part, whether the transformation of food systems threatens to squeeze smallholders out of the system altogether or just to prevent them from taking part in the most lucrative opportunities, thus imparting distributional effects. The more permeability between traditional and organized chains in the domestic food system, the lower the barriers to participation for smallholders (Figure 5). The differences between traditional and specialized wholesalers with respect to prices and standards will affect farmers' decisions on where to market their produce, along with other aspects of the point of sale that are explored below. There is evidence that specialized wholesalers can offer higher prices than traditional ones for produce that meets their standards (e.g. Schwentesius and Angel Gomez, 2002), but as discussed below, the cost of complying with standards can eat into this price differential. Furthermore, the price differential may decrease over time as more and more producers in a location are capable of meeting exacting standards.

In Asia, many farmers are participating in organized retail chains serving domestic supermarkets without knowing it. Supermarkets in China and India have been shown to procure at least some fresh produce from the traditional wholesale market (Huang *et al.*, 2008; Singh, 2008). In fact, smallholder fruit farmers in China's Shandong province were found to be supplying supermarkets in Russia and central and eastern Europe via the traditional marketing system (Huang *et al.*, 2008). To date, many Asian supermarkets have been able to scale up quickly by sourcing in traditional chains. This has been an effective strategy because there have not been major safety scares, or they have been sufficiently downplayed to avoid scandal. In China there has been a recent public backlash against the shortcomings of the public safety regulation system (Barboza, 2007). In the near future, retailers may come under greater pressure to introduce traceability into their sourcing systems.

Formalization of procurement and marketing

As the organization of retail leads to the specialization of wholesale, there is an expansion of formalized procurement systems designed to improve the efficiency of procurement. Formalized procurement systems facilitate the transmission of information upstream and downstream, allowing for differentiation of products based on quality and safety, and reducing the costs of coordination between buyers and sellers (Barry *et al.*, 1992; Pingali *et al.*, 2007). Procurement models are changing as marketing systems shift from traditional to structured to modern. Transactions in traditional markets are characterized as being 'spot' in nature, although relationships between farmers and traders are likely to be important in any market. Traditional markets may be regulated by institutions, such as government commodity boards, which require farmers to sell in certain channels (Jayne, 2008). Above all, traditional markets are characterized by informality, with farmers bearing many of the costs associated with poor market performance. Farmers may be charged irregular fees by various intermediaries, for example, and traders may depress prices through collusion (Omiti *et al.*, 2008; Singh, 2008).

Structured markets are characterized by more rules and regulations set by government overseers, although not necessarily by heavy-handed direct involvement. The Agricultural Produce Marketing Committee (APMC) system in India is an example of a well-regulated auction where farmers can expect certain basic rules of engagement. Rules are meant to standardize transactions and make the market a fairer place to do business (Singh, 2008). Governments may publish price information to reduce farmers' information asymmetry. To some extent, farmers can access channels for arbitration when they feel that regulations have been violated. Traditional and structured markets often contain many intermediaries. When there are more intermediaries in the chain linking producers and consumers, each intermediary earns lower margins, and the overall marketing efficiency of the chain is lower (efficiency is based on the difference between producer and retail prices). Farmers and consumers lose when markets are inefficient, so, in a competitive market, both stand to gain if margins are reduced.

Modern supply chains are characterized, above all, by coordination, which usually reflects some pre-arranged agreement of the price and non-price terms of a transaction. Fewer intermediaries are found in modern chains, and upstream and downstream linkages are tighter. Improved coordination, horizontally and vertically, is an effective strategy for reducing transaction costs in modern chains. Costs associated with poor transmission of information are certainly lowered. More efficient consumer response systems reduce costs associated with the bullwhip effect, which arises from delayed transfer of information about stocks from retailers to wholesalers, traders and producers (Fransoo and Wouters, 2000). When there is a disease outbreak in a food system with improved traceability, responses can be more efficiently targeted to the source of the outbreak, reducing total losses (McKean, 2001). Product traceability provides a mechanism for retailers to gain a competitive advantage on the basis of specific product attributes and for producers to gain price recognition for providing these attributes (Opara, 2003). At the same time, improved coordination introduces new transaction costs along the chain, which can diminish or even completely offset the gains from coordination (Pingali *et al.*, 2007). There are fixed, direct costs associated with building the necessary infrastructure for product differentiation and traceability. Initializing and developing more formal relationships between buyers and sellers can be a costly process, as buyers and sellers must be matched and then must negotiate terms of sale.

It is not undifferentiated commodities but products, characterized by specific attributes such as size, shape and colour, that flow through modern supply chains. Modern retailers and specialized wholesalers often use preferred supplier lists to lower their transaction costs. The buyers decide which farmers appear on the preferred supplier lists, and they often prefer larger farmers because of the fixed costs of transacting with each supplier (Pingali *et al.*, 2007). Buyers may pass these transaction costs on to smallholder farmers or they may otherwise exclude them from contract opportunities by opting to procure from larger suppliers. There are also many reasons for and evidence of modern retailers and processors procuring regularly from smallholders. Smallholder farmers may be able to provide better quality assurance at a lower cost of enforcement. Thai Fresh United relies on many small producers, who use labour-intensive techniques, to supply high quality herbs, spices, vegetables and fruits that meet their company's strict requirements (Boselie *et al.*, 2003). Purchasers can also diversify their supply base and stabilize their supply stream by sourcing from smallholder farmers (Kirsten and Sartorius, 2002; Dries *et al.*, 2004). When there are not enough larger farmers willing or able to produce the required volume and quality, buyers must turn to smallholders.

Major retailers and processors with a wide sourcing base have the ability to move goods among domestic markets, regionally or internationally at low cost given their economies of scale. Therefore, they can afford to be more price sensitive than firms with a smaller procurement base and to be more mobile in their procurement practices. Given the potential mobility of retailers with regards to sourcing, it is unrealistic to expect them or their procurement agents to pay prices that are higher than they would pay elsewhere, or to tolerate high transaction costs in the procurement process. For this reason, sustainable inclusion of smallholders in modern chains must rely on cost-effective models for bridging the scale discrepancies between individual smallholders and modern buyers.

As organized retailers and wholesalers capture a growing market share, they must assume the challenge of expanding procurement. Some must buy outside of their managed chains to fill orders (Singh, 2008). Farmers and intermediaries who sell from traditional into formal chains must demonstrate, or be accountable for, the quality and safety of their produce. There are incentives for independent traders to bring produce into organized chains from the traditional market in order to take advantage of higher margins and the stability of organized outlets. These intermediaries can assume the quality risk themselves (by procuring from the traditional marketplace without paying a premium, sorting as necessary, and reselling) or they can pass the quality risk on to producers (by offering a price premium and imposing informal quality standards).

3. Implications and impacts for smallholder agriculture

A useful way for policy makers to conceptualize the impacts of current and future trends in the organization of food systems on smallholder farmers is to identify the incentives, opportunities and constraints they pose, now and in the future, from the small farmer perspective. The transformation of consumption and restructuring of supply chains have created new market opportunities for many agricultural producers. These opportunities may still hold fringe status relative to traditional marketing systems, but they are typically growing much faster. Rapid shifts to capital and knowledge intensive production technologies and the importance of scale are likely to exert pressure on smallholders to adjust, though they may lack the means to do so. If widespread 'exclusion' is, in fact, being observed, it could foretell difficult times for smallholder farmers as dynamic, high-value market outlets take over more of the market share. The speed of adjustment is important because it impacts the ability of smallholders to adapt to new changes and the possibility for an 'orderly exit' from agriculture. The latter involves building the appropriate human capital for employment in the off- or non-farm rural sectors or for migration. It is important to note that the migration of rural poor to urban areas does not reduce poverty but rather transfers it to cities (Ravallion *et al.*, 2007).

Smallholder opportunities should arise from areas where smallholders are able to sustain a competitive advantage, such as low supervision costs when household labour is used. Similarly, constraints relate to aspects of production and marketing that prevent small farmers from exploiting their competitive advantages, such as financial capital or lack of experience coordinating with buyers. Once situation-specific opportunities have been identified, policy makers can facilitate efforts to pursue opportunities and overcome constraints. These efforts need not require chain-specific investments, but rather broader steps and strategic investments to create the conditions that encourage others to invest, and also to create alternative opportunities for income generation by smallholders outside of production.

Here we lay out, key, direct, and indirect pathways by which the restructuring of food systems impacts both smallholder farmers and the rural poor. We discuss concepts and evidence surrounding participation, terms of sale received, costs of participation and their broader impacts on production systems and interactions with the rural non-farm economy. We focus on the links between the reorganization of food systems and smallholder agriculture, while drawing on the work of others who have emphasized the importance of smallholder agriculture for rural poverty reduction. In an economy where there are market failures, household consumption and production decisions are not separable (Singh *et al.*, 1986), and the transformation of food systems affects smallholders' production and consumption decisions.

3.1 Why the small farm focus?

As of 2007, the world's population shifted from one that was mostly rural to one that is mostly urban, but for developing countries the majority of the population is still rural. Since poverty rates in rural areas exceed those in urban areas, most of the world's poor are still found in rural areas (World Bank, 2008). Although most rural households have diversified income sources, the majority of rural poor earn their income in agriculture (Davis *et al.*, 2007). Even though farming households are highly diversified with respect to income sources, agricultural production and marketing remain important determinants of household welfare, especially those with lower incomes. Amongst farming households, smallholders are more likely to be poor than those with larger land holdings (Davis *et al.*, 2007). Smallholder farmers, thus, comprise a substantial part of the rural poor demographic and are therefore a logical entry point for an analysis of how the reorganization of food systems impacts rural poverty.

Furthermore, smallholders form the 'structural backbone' of the rural economy because of their linkages with small-scale input and service providers, traders, backyard processors and hired labourers (Ashley and Maxwell, 2001). Smallholder productivity and income gains are translated into demand for labour-intensive consumption goods produced in rural areas and also into investment in non-farm rural activities, thus creating multiplier effects in rural economies. Smallholders are also of interest because, in many instances, they are known to use land more productively than farmers with larger landholdings (Berry and Cline, 1979; Helfand and Levine, 2004). The inverse productivity relationship underscores the importance of pursuing growth in the smallholder sector, since gains can be shared broadly. Furthermore, it is posited that the transformation of food systems creates new opportunities for smallholders arising from aspects of production that are not scale sensitive. For instance, labour market imperfections, which result in a low opportunity cost of household labour, allows for cost effective supervision of production systems (Heltberg, 1998).

Despite the strong linkages between the small farm sector and rural poverty, it is important to understand the distinction between farms and farming households. Rural households show a high degree of diversification in their activities, with income from agriculture and livestock supplemented by farm and non-farm wages, remittances and income from small enterprises. However, the vast majority of rural households in developing countries have some form of participation in agriculture while, despite diversification, farm income is the backbone of the income structure of poorer households (Anriquez and Stamoulis, 2007). A broad look at the changing structure of food systems, and rural economies, must incorporate the different modalities by which households can be affected by a changing rural economy. Farming is but one entry point, albeit a very important one. To date, there has been little broad-based analytical work addressing the changing income patterns of rural households and composition of rural economies. The analysis shows that expansion of modern forms of retail organization has wider impacts in rural economies, on both farm and non-farm activities (Reardon *et al.*, 2007).

Transformation of production systems

As retail becomes more organized, wholesale more specialized, and procurement more formalized, the management and composition of production systems is transforming. As a general rule, production systems are becoming more commercialized in developing countries. Commercialized systems are characterized by specialization at the farm level, greater dependence on purchased inputs and more marketing of outputs (Pingali, 1997). Typically, commercialized systems use more labour and inputs per unit of land than subsistence systems. Input use remains low in production systems that are not closely linked with markets (Heisey and Mwangi, 1997; Omiti *et al.*, 2008; Tobgay and McCullough, 2008). As urban centres demand higher value products, and as market structures respond to urban demand, we observe diversification of cropping systems at the meso- and macro-levels, even as they become more specialized at the farm level to take advantage of economies of scale. In Bhutan, areas closer to road points exhibit more market-oriented specialization of their cropping systems and a greater likelihood of participating in output markets (Tobgay and McCullough, 2008).

Particularly in areas where land holdings are small and arable land is limited, smallholders begin to specialize in higher value enterprises, such as horticulture and livestock, as opposed to lower value cereals. At the meso-level, diversification into higher value cropping systems is limited by agroclimatic potential, water resource development and the strength of market linkages. At the farm level, assets, technical know-how and labour availability can limit diversification into higher value crops. In agro-climatically less favoured areas, there are fewer opportunities to produce higher value products for modern chains. Extensive livestock production offers some potential, as does production of lower value, non-perishable raw materials for processing. Biofuels markets could provide high return opportunities in some places where other options are not available. Farmers in less favoured areas are more likely to be competitive in diversified and mixed-livestock systems (Cassman, 1999). Without specific R&D efforts targeting less favoured areas, however, prospects will remain limited. Major constraints include lack of irrigation, pests, poor soil structures and nutrient limitations. Interventions to alleviate these constraints may focus on breeding and variety development, improving best practices for field management and capacity building with technology transfer.

In many agricultural economies at the low end of the development process, agricultural inputs are expensive relative to global prices and, as a result, underutilized. Costliness of inputs arises partly from underdevelopment of infrastructure and underinvestment in institutions (Jayne, 2008). When input prices are too high relative to farm gate output prices, it simply does not make sense for farmers to purchase modern inputs. Furthermore, input providers are not well regulated, and many farmers bear the costs of dubious quality seeds, fertilizer and other agrichemicals (e.g. Crawford *et al.*, 2003; Omiti, 2008; Tobgay and McCullough, 2008). Increasing input use is only profitable to the extent that productivity gains offset the costs of inputs. In many Asian and Latin American countries, input subsidy programmes played a historic role in raising crop yields (Falcon *et al.*, 1983). Most of these programmes were phased out as economies were liberalized, with government agencies now participating less directly in input provision. Similarly, government programmes for agricultural credit, extension, marketing and germplasm development have been scaled back across the developing world as governments have come into compliance with international trade agreements and unsustainable budget imbalances (Jayne, 2008).

The void in agricultural support services has been at least partly filled by the private sector. Input manufacturers and retailers have long played a part in providing agricultural extension. But now buyers are playing a more prominent role in the provision of agricultural services. This includes fertilizer and chemicals, technical assistance and the provision of seeds. Because of their scale, modern buyers can leverage government programmes to subsidize extension or irrigation investments, for instance, receiving bulk payments in exchange for administering services and/or subsidies to farmers. By providing inputs and services for free or below retail prices, buyers can improve their control over production processes while producers improve their access to services.

Upstream from producers, there has been global consolidation in manufacturing of and R&D for key agricultural inputs (Kimle and Hayenga, 1993). This reinforces the hourglass structure, in which a growing number of producers find themselves sandwiched between large, multinational firms who control input manufacturing on one side, and processing and retail on the other (Pingali *et al.*, 2007). Consolidation in input manufacturing is a result of the considerable economies of scale involved on the R&D side. The example of consolidation that took place in the seed industry is telling. Crop by crop, as the industry has advanced from the pre-industrial to the mature stage, private firms replace state agencies in dominating key germplasm R&D activities. The private sector invests more and more in R&D where intellectual property laws are more secure. Even in mature seed sectors, though, the state's role remains important in regulating the sector and providing complementary public goods (see Morris *et al.*, 1998). The same consolidation is taking place with respect to other agricultural inputs, and the state's role in regulating the input manufacturing sector is underscored. Another role of the state is to facilitate R&D for agriculture in less favoured areas, which can bring high returns but may nevertheless be ignored by the private sector (Fan and Chan-Kang, 2004).

3.2 Formalizing terms of sale

The ability of smallholders to sustain participation in organized chains depends on how the terms of sale and cost of participation for the modern chain compare with those for traditional alternatives. Expected returns are impacted by prices and their stability, cost of transporting goods, rates of rejection and timing of payment. Terms of sale relate to the price used for a transaction, but also when, where and how the transaction takes place. They dictate what product changes hands, what standards it must meet and how testing will be conducted.

Terms may depend entirely on the bargaining skills of the parties involved; they may be governed by the regulations of a marketplace; or they may be agreed upon in advance and specified in a written contract. As food systems transform from traditional to structured to modern, there is a shift from the former to the latter. Terms of sale matter because they determine how incentives, risk and marketing costs are distributed between the buyer and the seller. While price and quantity sold can be tracked rather easily, costs (particularly transaction costs) are more difficult to measure and differ greatly between farmers and contexts. Differences in negotiation skills, experience and affiliations can lead to differences in terms of sale (Pingali *et al.*, 2007).

Modern procurement systems may offer participants higher prices, but they also introduce new risks and costs. With fewer and more powerful buyers, farmers have reduced power for negotiation (Gibbon, 2003; Timmer, 2008). Some farmers reported that modern chains offered lower prices but more stability. Others perceived them as offering higher prices but being more risky due to a lack of transparency in quality assessment or price setting. In many transactions, the party that bore more risk (i.e. through price variability) also garnered incentives for doing so (i.e. a higher share of marketing margins). In Bhutan, for example, citrus exports to India constitute one of the more modern chains. The model relies on intermediaries, who receive advance finance from exporters, to procure oranges from smallholders (Tobgay and McCullough, 2008). As soon as citrus trees blossom, collectors provide advance credit to their producers in exchange for assured access to their orange harvests at a fixed price, determined by the blossoms. The collector oversees the harvest, transport and marketing of the products and bears all associated risks. In the absence of such arrangements, smallholders would likely be deterred altogether from citrus marketing by the labour costs of harvest and the risk of product spoilage due to poor road and market infrastructure.

In general, perceptions of fairness regarding terms of sale have a lot to do with prevailing conditions in surrounding markets. True impacts can only be evaluated after multiple years of repeated participation in a chain, but judgements are often made much sooner. In general, modern chains stabilize inter-annual risk related to price and market instability while introducing new risks related to higher costs of participation and more exacting requirements. Problems arose when risks were delinked from rewards. Sellers, in particular, who bore more risk with less reward, felt they had received asymmetric terms of sale. The perception of unfairness in terms of sale most commonly arose from buyers' quality assessment, requirements for chain-specific investments and misconceptions between buyers and sellers that led to side-selling.

For instance, when a farmer becomes party to a contract with a fixed price, he or she bears the risk of a price increase while the buyer bears the risk of a price decrease. With a floating market price, both parties share all price risk when information is symmetric. The specific way in which quality standards are enforced also affects the distribution of risks and rewards. An agreement may be designed to penalize a seller for failing to meet standards, either with a price cut or outright rejection. If quality assessment takes place at the point of sale, and the seller assumes transport costs, the seller bears a disproportionate risk from crop rejection. In India, contracts drafted by McCain and Frito Lay for potato growers allowed buyers to reject produce for any reason, despite the fact that producers were obligated to pay for transportation costs to the drop-off facility (Singh, 2008). Producers felt this placed too much risk on them and complained that the quality inspection process was not fair or transparent. Similarly, in Kazakhstan, cotton farmers complained that the buyer, who also performed quality assessment, had incentives to underestimate quality so they could pay lower prices (Swinnen, 2005).

When suppliers provide inputs and a fixed price, they may then offer a price that is lower than the average market price. From their own perspective, in this model, buyers bear a price risk and a default risk. Sellers, particularly those without a good relationship with the buyer, may not understand the logic behind price setting, and, on seeing a better price elsewhere, may choose to side-sell into a different channel. In Kenya, many dairy farmers were bound to sell to their cooperatives in return for the technical assistance they received. They often sold at least a portion of their milk production outside of the cooperative, though, to illegal hawkers, who offered higher prices (Omiti *et al.*, 2008). Unfairness in terms, or perceptions of unfairness, arise from a lack of transparency in the process of formulating and enforcing terms of sale. Interventions should be targeted towards improving understanding while opening channels for conflict resolution. Collectivizing farmers' bargaining holds the promise of improving terms of sale from the smallholder perspective. Tools for doing this are explored below, along with their costs of implementation.

Contracts

Contract farming is a mechanism for vertical coordination that is growing in popularity in modern chains. Contracts usually involve advance agreement between producers and purchasers on some or all of four parameters: price, quality, quantity (or acreage) and time of delivery (Singh, 2002). Specific contract terms and arrangements determine how the parties involved share the benefits, costs and risks of coordination. These may deal with timing of payment; mechanisms for setting price; provision of services and inputs; documentation requirements; quality and quantity produced; arrangements for assessing quality; and mechanisms for settling disputes and enforcing agreements. When contracts fix output prices in advance, they may allow farmers to produce risky high-value, perishable crops that they otherwise would avoid because they are prone to a price glut. These arrangements can also help to ensure a reliable supply for companies that have made sub-sector specific investments (Simmons *et al.*, 2005).

In labour markets, farm owners and wage labourers choose to enter fixed labour contracts because of shortcomings in labour markets. The shortcomings arose from seasonal risk in the demand and supply of labour (Bardhan, 1983) and from difficulties in monitoring casual labourers in tasks like irrigation and input application (Eswaran and Kotwal, 1985). Similarly, buyers and sellers may choose to enter fixed marketing contracts in order to overcome risk and uncertainty in spot markets. These risks of spot markets are similar to those of casual wage labour markets, arising from seasonal variability in supply and demand (resulting risk of shortage and surplus) and from the need to assure quality in the absence of perfect monitoring.

From the buyer's perspective, the cost of procuring via contracts includes transaction costs arising from the design and implementation of a contracting system (Pingali *et al.*, 2007). Managerial costs, along with capital investments in facilities, are involved. Retailers and processors who procure through contracts must also plan for the costs of abiding by contract terms, which may involve providing inputs at fixed or below-market cost, providing technical assistance and providing credit. More costs result from carrying out transactions and enforcing contracts, including testing product quality and safety, and arbitration where necessary. Many of these costs have fixed, per farmer components, which buyers can cut by targeting larger producers.

As with terms of sale in a non-contract transaction, specific contract terms will determine the extent to which small farmers can share in the benefits from vertical coordination because they allocate risks between interested parties, such as price and market risk, crop failure risk and the risk that a contractual party defects. Important contract terms include timing of payment; mechanisms for setting price; provision of services and inputs to suppliers; demands on documentation, timing, quality and quantity; arrangements for assessing quality; and mechanisms for settling contract disputes and enforcing agreements. A favourable legal and institutional environment helps contracts to be fairer for small farmers. The Model APMC Act in India, for example, requires contracts to be registered with a local authority and includes provisions on contracts, liabilities, asset indemnity and dispute resolution (Singh, 2008). Direct contract relationships between producers and corporations proved to be more beneficial for small farmers in Punjab than state-sponsored contracts. They resulted in better delivery of extension services and more reliable purchase of commodities (Kumar, 2005).

Contracts can help smallholder farmers access key inputs and services that may otherwise constrain production. The contract itself gives buyers some assurance that they will capture the benefit stream from investing resources in producers. Buyers often provide inputs to farmers with whom they are contracting at below-market prices or at cost, or they may provide technical support and extension services, often of better quality than publicly provided extension services (e.g. Kumar, 2005). Contracts can also facilitate access to credit. In Kazakhstan, credit was the primary reason that smallholder cotton farmers entered contractual arrangements (Swinnen, 2005). In Lithuania, the only source of credit for small dairy farmers was through buyers procuring with contracts. Dairy purchasers in Poland offered credit along with extension services and inputs, and farmer participation was very high in return (Dries and Swinnen, 2008). Supermarkets have offered similar provisions to small farmers, via contracts, throughout eastern Europe and central America. Farm assistance programmes created for contract farmers have been replicated by other companies and by state agencies because of their success (Swinnen, 2005).

Farmers have likely benefited from contractual arrangements in a number of instances, but it is difficult to attribute benefits to participation in the contract itself as opposed to participation in the chain. Benefits arising from contract farming often spill over to participants' non-contract fields and to neighbouring farmers. In central and eastern Europe and the former Soviet Union, contract farmers enjoyed higher productivity with lower risk on their non-contract crops (Swinnen, 2005).

However, there is also abundant evidence that smallholder farmers are excluded from entering formal contracting arrangements. In India, the contract farming system favours larger farmers at the expense of small producers, very few of whom are participating in contract farming (Kumar, 2005; Singh, 2008). In central and eastern Europe, it is more common for farming corporations to enter contractual agreements than it is for small farmers (Swinnen, 2005). Although contract farming has risen to the point of including 9% of farmers in Suphan Buri, Thailand, very small farmers are much less likely to enter contracts (Dawe, 2005). They are also less likely to receive favourable terms, such as a fixed output price, and therefore bear more risk.

Farmer organizations

One possible method for small farmers to overcome some of their size disadvantages is to form production and/or marketing groups. By joining together in the name of common production and marketing interests, small farmers can increase their effective size and bargain for more favourable terms. Cooperation can increase bargaining power, allow for economies of scale, and lower marketing and negotiation costs by pooling negotiation efforts. Cooperative marketing can ease supply constraints faced by individual farmers, allowing them to meet buyers' orders year-round, where production systems permit. Internal incentives can be provided for farmers who fill off-peak orders. When agricultural systems are dominated by smallholders, farmer organizations have been designed in order to supply large buyers who have no other options for procurement. In India, Mahagrapes successfully arose as an export-oriented umbrella marketing organization for several cooperatives of smallholder grape producers (Narrod *et al.*, 2008).

When buyers require investments, farmer organizations may offer a cost-effective way of upgrading through pooling of investments. By organizing, smallholders can access information and share knowledge more easily, decreasing their search and information costs. There is a strong tradition of farm cooperatives in the Netherlands, which have served as a farmer safety net and helped to raise productivity. Now Dutch cooperatives are assuming many more roles, including innovation and direct involvement with consumers (Bijman and Hendrikse, 2003).

Farmer organizations are not a panacea, as they can be very costly to set up and maintain. Efforts to design and start an organization in one place are not necessarily replicable because management structures are so contextualized. Group decision-making can be costly, and, in some cases, the success of organizations is unduly dependent on the charisma, intelligence and altruism of one leader. An informal survey of supermarket procurement officers worldwide suggested that retailers have negative associations with procurement from farmer groups, stating that they can be difficult to work with, unreliable and inexperienced (Reardon and Hopkins, 2006).

Lowering marketing costs: other strategies

Beyond the widely discussed cooperative, there are other models for achieving economies of scale through coordination between farmers. Large farmers, for instance, can serve as intermediaries between smallholders and supermarkets by subcontracting for some of their production needs (e.g. in Honduras) (Lundy *et al.*, 2006; Meijer *et al.*, 2008). Different forms of tenant farming (e.g. exchange of labour for a portion of harvest) and reverse tenancy (e.g. leasing of land management to a larger operator in exchange for rent) have long been in practice to solve various inefficiencies in factor markets, particularly for land and labour.

Geographic clustering by product has been put forward as a way to economize on sub-sector specific investments in production and post-harvest infrastructure. This strategy may offer some promise, but picking sub-sectors that will retain price stability is notoriously difficult. Mistakes have been made in the past, with farmers suffering the effects of a price glut due to overproduction while struggling to repay debt on specific assets (Shepherd, 2007).

Costs of participation

Modern chains often dictate production methods and may or may not facilitate support for production systems via technical assistance and input provision. Production costs are likely to be higher in modern chains, which are more demanding than traditional chains. Evaluation of explicit costs and returns, and less explicit transaction costs associated with maintaining and enforcing agreements, will dictate smallholders' competitiveness in and preferences for different chains. Because the same characteristics that allow a farmer to supply a high value market will also influence the farmer's income regardless of market, higher incomes observed in a chain can result from either the chain's characteristics or the farmer's, or some combination of the two (Sadoulet and De Janvry, 1995). Sustained participation over time is a good indication that a chain is profitable for participants compared to other options. If farmers are required to make specific investments, though, especially in specialized assets and equipment with low resale value and convertibility, ex post continued participation might reflect investment irreversibility and sunk costs rather than satisfaction. Farmers who make specific investments in order to participate in a chain must bear the risk of the buyer defaulting (Gow and Swinnen, 1998). It is important to have watchdog organizations or institutions accessible so that they can voice their complaints and pursue arbitration when they feel they are being exploited.

Apart from the ability to specialize in specific crops, farmers selling into modern chains must be able to meet their more exacting quality and safety requirements. Complying with private and public standards has implications for on-farm production systems. It may require investments in capital equipment, such as post-harvest storage facilities, or a system for preventing contamination of fields with household waste water (Narrod *et al.*, 2008). In Kenya, smallholders who were supplying fresh green bean export chains switched to chains for processed beans once stringent quality and safety standards were introduced into the fresh green bean chain (Narrod *et al.*, 2008). It is very difficult to meet high quality standards for horticultural crops without an irrigation system, which allows for efficient application of inputs (Rosegrant *et al.*, 2002). In the heavily groundwater-dependent Indian state of Gujarat, McCain informally required its potato suppliers to use efficient irrigation systems, citing concern about sustainability of water use as the reason (Singh, 2008).

In order to encourage better quality, modern buyers urge their suppliers to adopt specific management practices regarding varieties used, planting, fertilizer and pest management, and harvest. This was observed in virtually all case studies. Retailers have been known to request their suppliers to adopt integrated pest management to reduce the prevalence of pesticide residues in final products. To encourage uniformity of produce, processing firms may dictate specific dimensions for seed bed height and width and planting date (e.g. Singh, 2008). In general, modern chains will be more closely linked with consumer demands (or processor requirements) since they will have in place mechanisms to transmit information and incentive systems upstream to reward compliance. In countries where quality requirements for traditional domestic systems differ greatly from those in developed countries, production practices can differ drastically between fields with crops produced for export and with the same crops produced for the domestic market (e.g. Narrod *et al.*, 2008).

Smallholder participation is limited

Ultimately, not all farmers have the option of supplying all markets. From the options available to them, farmers will choose the ones that bring the most expected returns to the household. It is rather easy to observe whether or not smallholder farmers are participating in a given marketing chain. But non-participation is not the same as exclusion, since it can also arise from a farmer's decision not to participate because a different option is preferred. Distinguishing between these two types of non-participating farm households can be difficult without targeted surveys at the household level, but confusing them can lead one to erroneous conclusions. There is evidence from all areas of the world of smallholders participating in many different types of modern chains, both domestic and for export, with contracts or without, as part of producer organizations or independently. However, it is very difficult to assess the extent of participation because most studies adopt a case study approach, tracing a particular retailer or producer group, or targeting a location because participation is known to take place there.

This case-based approach is necessary for identifying and assessing emerging trends, but it is not good for estimating their extent. Evidence from central and eastern Europe suggests that smallholder inclusion is robust in areas where most landholdings are uniformly small (Swinnen, 2002). Where smallholders are part of a dualistic system with the presence of large landholders, modern buyers show a preference for procuring from large farmers. Evidence from Latin America and Africa supports this hypothesis (Berdegue *et al.*, 2005; Reardon *et al.*, 2008). Modern buyers have been known to develop mechanisms for procuring from smallholders because there is no one else to procure from (Narrod *et al.*, 2008) or for public relations purposes. It is difficult to know the extent to which public relations incentives have motivated smallholder inclusion, but such incentives are likely to be limited in nature and short-lived.

Scale mismatch is perhaps the most common constraint to smallholder participation in modern chains. Individual smallholders have limited ability to negotiate and bargain for beneficial price and non-price terms from major retailers and processors on the output side and major multinational manufacturers on the input side (Vorley, 2003; Pingali *et al.*, 2007;). Smallholder farmers can be excluded from preferred supplier lists or contract-based marketing channels because buyers specify a minimum cut-off acreage or product volume that exceeds their capacity, given finite land holdings. It is much more likely, though, that smallholders will be excluded *de facto* because of fixed costs involved with participating in modern chains. Buyers may require specific on-farm investments in assets (e.g. irrigation systems) that are not profitable for small-scale producers. In Honduras, for example, Hortifruti's regional specialized wholesale arm required farmers to pay for their own costs of supervision. The costs started at 1000 US$ per year in each farmer's first year but were reduced to 500 US$ in subsequent years (Meijer *et al.*, 2008). In the Shandong province of China, farm size and household assets had little effect on which marketing channel apple and grape farmers took part in, but strict quality and safety standards were not present (Huang *et al.*, 2008).

Another key barrier to smallholder participation in modern, high-value marketing opportunities arises from business orientation. Identifying, solidifying and exploiting opportunities to sell to modern buyers requires a certain entrepreneurial quality. Some farmers have fewer opportunities to develop their managerial human capital (Bingen *et al.*, 2003). Because transaction costs associated with joining modern chains are likely to be large at the start and decline with time and experience, policies directed at lowering initial barriers to entry for farmers who are otherwise competitive are likely to be effective at facilitating the inclusion of smallholder farmers. Interventions aimed at reducing uncertainty surrounding new outlets and improving advocacy tools available to small farmers are likely to reduce one-time transaction costs associated with entering modern chains. Negotiation costs can be reduced with capacity building and legal assistance with forming agreements. Search and information costs can be alleviated with the expansion of market information systems that make marketing outlets and their terms more explicitly known and allow farmers to compare different outlets (Pingali *et al.*, 2007).

Empirically, it is difficult to distinguish between buyer exclusion and farmer self-exclusion because the observed outcome (non-participation) looks the same. If modern chains involve higher costs but bring higher returns, and smallholder farmers are resource constrained, buyer exclusion is probably a more common cause of non-participation than self-exclusion. Several examples uncovered in this research suggest instances where smallholders were capable of participating in modern chains but chose not to. In India, for instance, farmers who were selling potatoes under contract to modern processors chose only to sell about 50% of their output to the processor. They were qualified to be preferred suppliers, but in order to hedge the risks of full participation, they only committed a portion of their land holdings to the contract (Singh, 2008). With time, the costs associated with introducing formal coordination into transactions should fall as transactions are repeated between the same parties, who acquire experience and build trust in the process (Rademakers, 2000).

It is important to note that even farmers who sell into modern chains also sell into traditional (or structured) wholesale markets at some point, even for the same crops that are being sold to modern buyers (McCullough and Pingali, 2008). Improving the performance of traditional markets benefits everyone, as alternative procurers must compete with the traditional system. The issue of smallholder participation in modern chains has limited poverty reduction implications for farmers who are not already participating in markets. It is important to note that, for many smallholders, participation in even traditional markets remains a more pressing concern. Household surveys conducted across eastern and southern Africa suggest that the majority of rural households do not sell any grain but buy it regularly (Jayne, 2008). For many of these households, the most important source of 'income' is household production that is consumed at home. By improving productivity and reducing marketing costs, households can divert more labour to cash-earning activities.

3.3 Interactions between the transformation of food systems and the rural non-farm economy

With income growth, increased opportunities are available for off-farm employment, and more pressure is placed on labour markets, resulting in rising wage rates which also affect seasonal agricultural labourers. The diversification of production systems out of staple crops and into higher value products is another characteristic associated with the transformation of food systems. Higher value crops, such as horticulture and livestock, often require more labour input. High-value exports from Senegal had a poverty reduction effect through labour markets rather than smallholder participation (Maertens and Swinnen, 2006). Households' willingness and ability to grow higher value crops is impacted by the availability of labour within the household and the predominant wage rates for hiring-in labour. In the Shandong province of China, greater household participation in off-farm income earning activities was associated with lower participation in fruit production. Fruit production and off-farm employment were seen as competing demands for the time of household members. Those who were involved in off-farm employment faced higher opportunity costs on their time and were less likely to turn to apple and grape farming (Huang *et al.*, 2008). Some evidence suggests that off-farm income is correlated with agricultural input use, which is consistent with the hypothesis that it eases credit constraints (Davis *et al.*, 2007).

As food systems transform, so will the non-agricultural prospects of smallholder households. Households may choose to pursue off-farm work in agricultural processing, other manufacturing, agricultural labour markets or through migration to other places. Little is known about the impacts of changing food systems on the broader rural economy (Reardon *et al.*, 2007). Impacts will arise from subsidiaries of supermarkets moving into small towns and rural areas to sell food and other consumer items, thus knocking out local small-scale retailers and businesses. Since smallholder agriculture constitutes the backbone of the rural economy, marginalization of smallholders will likely have net labour effects in rural areas (Anriquez and Stamoulis, 2007). It is natural for populations to move from farms to cities as an economy grows and the relative importance of the agricultural sector falls. But agriculture is important for supporting households until better opportunities emerge in other sectors, preventing premature exit. Many African countries have much higher urbanization rates than they would if agriculture was more productive (Jayne, 2008). Managing the 'push' out of agriculture would help alleviate social problems arising from growth rates of urban areas. For those without prospects for migrating or working in other sectors, agriculture is the only hope.

4. Way forward

Because of organizational changes in food systems, smallholders now face many new opportunities to benefit from rapidly growing market segments. Modern chains are capable of lowering the risks of participating in higher value markets while transmitting rewards for meeting quality and safety standards. Because of the high costs of participating in these marketing chains, smallholders risk being excluded from a lucrative market segment. In many transforming economies with modernizing food systems, modern chains account for a substantial share of food retail. In these countries, if retailers are unable to procure through traditional channels, they will form a separate, vertically coordinated procurement system that competes with the traditional system. This can be observed in many countries in Latin America. In this case, the risk to smallholders is that, unless they can link into the modern procurement system, they will be relegated to a low value, shrinking market segment. Throughout Asia, traditional markets have continued to supply the modern retail sector, which itself has captured a limited share of food retail. While rising food safety concerns amongst Asian consumers could threaten ties between traditional wholesalers and modern retailers, for now Asia's smallholders appear to be well linked in with domestic urban markets. In many parts of Africa, the danger is not that growth in modern retail has captured or will soon capture a substantial share of food retail. Furthermore, lucrative opportunities to link with high-value export markets are quite limited in scope. The real danger is that smallholders in remote areas are excluded from agricultural marketing altogether due to high transportation and transaction costs and the widespread availability of cheap imports.

The transformation of food systems presents a set of problems that vary drastically between countries, based on characteristics of the food system, the place or the households involved. Different countries will prioritize problems differently depending on the context. There is no one policy response, but a common objective between all situations is to see smallholders through the transition, in recognition of their importance for rural poverty. Facilitating the transformation ultimately boils down to a three-pronged policy approach:
1. facilitating the inclusion of smallholders in modern chains by reducing costs of participation;
2. upgrading traditional marketing systems;
3. supporting those who cannot supply traditional markets with social safety nets.

4.1 Policy responses

We explore in greater depth what each initiative entails and outline a role for governments, the private sector and civil society to play in facilitating the transformation for smallholder farmers (Table 3). In general, identifying and pursuing appropriate policy responses to the transformation of food systems presents several challenges. The transformation of food systems is an unwieldy phenomenon, often spilling out of the traditional policy space of Ministries of Agriculture. Responses at different phases of the food chain must be coordinated, and it is not always clear which institutions and ministries can and should take the lead role. Responding to the transformation of food systems does not require drastic reforms. Relatively minor adjustments, beginning with

Table 3. Policy tools in traditional, modernizing and industrial food systems.

	Traditional	Modernizing	Industrialized
Consumption	Targeting chronic hunger victims	Promoting health and diet diversification	Minimize health burden of ageing and obese population
	Promoting diet diversification	Preventing obesity through consumer education	Internalizing public health costs of unhealthy foods
Retail	FDI[1] for retail and agribusiness	Public safety standards and regulation	Farmers markets
	Basic hygiene in traditional markets	FDI for retail	Certification for niche products
Processing	Expand value added through processing	Expand exports of processed goods	Downscaling 'safe' processing facilities
Wholesale	Upgrading traditional wholesalers	Upgrade traditional markets	Improved ICT for bypassing wholesalers
	Improving safety Reducing spoilage	Improved safety Better traceability	
Marketing	Transport infrastructure	Targeted incentives to source from smallholders	Improved farmers' tools for direct marketing
	Enabling marketing institutions and information systems	Improve marketing capacity Improve regulatory environment	
Production	Increase market orientation	Meeting standards	Phase out unsustainable subsidy schemes
	Investing in productivity	Improving productivity and input use efficiency	Sustain natural resource support systems
	Diversification out of low-value staples	Sustain natural resource support systems	
	Regulate input providers		

[1] FDI: foreign direct investment.

removing market distortions and maligned incentives, can be very effective. However, developing country institutions are rarely of the cross-cutting nature needed to face such problems. While specific policy interventions should be addressed at the right scale, political boundaries and political capital do not always correspond with the scales at which market interventions are needed (i.e. market-shed, watershed, etc.).

Most of the policy interventions described above are relevant for governments in developing countries that are going through the agricultural transformation process. While appropriate government policies are absolutely necessary for managing the transformation of food systems, there are nevertheless roles to be played by the private sector and civil society organizations. Multinational corporations who are building procurement programmes in developing countries should be discouraged from pursuing monopsonistic[3] procurement conditions through anti-competitive behaviour. When procurement practices are open and transparent, it is easier to monitor and regulate them. Private companies in retail and processing cannot be expected to save smallholder farmers while serving their business interests. Therefore, creative institutional innovations are welcome to align the interests of smallholders with those of the modern retailers and processors who are controlling a growing share of food retail.

---

[3] Monopsony is a market characterized by one buyer and many sellers.

Non-governmental organizations (NGOs) can play a key role in supporting smallholders through the transformation when government policies fall short. They can help link smallholders with modern chains by lowering the one-time, initial barriers to entry. NGOs can do this by building capacity, providing information and experience, and financing investments in assets. Furthermore, NGOs can monitor vulnerable groups who risk marginalization by the transformation process. They can flag problems and mobilize political capital for addressing problems. NGOs and socially-oriented businesses have been involved in developing markets, through certification programmes and direct trade, in building niche markets for products whose supply chains are socially responsible, which have benefited the participants although the scope is still limited.

Governments in developed countries bear the responsibility of promoting a balanced system of global trade. In many developed countries, domestic support systems for agriculture have been widely criticized because they are linked with price distortion and commodity dumping. A new form of protectionism is arising in some developed countries, with retailers being urged to label the 'food miles,' or physical distance travelled by all products on their shelves. Improving consumer access to information about the energy footprint of the products available to them is essential, but food miles labelling isolates only one stage in the supply chain, the transport stage, rather than the entire chain. Products that are produced by smallholders in developing countries are likely to travel a longer distance to get to retailers' shelves, but their carbon footprints may nevertheless be lower than those of locally produced alternatives due to differences in production technology.

4.2 Facilitating the inclusion of smallholders in modern chains

At the country level, opportunities for linking smallholders into modern chains are determined by the size of the domestic market, which is set by urbanization rates, the average income and the prevalence of a middle class. Potential for trade is set by macroeconomic conditions and trade policies. Overall governance influences the cost of doing business (which has a large impact on transaction costs), and the institutional setting (which affects the climate in which agricultural activities occur). In countries with a strong urban demand, good institutions supporting financial services and R&D, and good governance, it is less of a battle to link smallholders with consumers through organized chains.

Most modern chains involve higher costs of production along with fixed investments. Households with asset constraints may not be able to overcome initial hurdles associated with entering a chain. Off-farm sources of income and household ownership of fixed assets most likely improve the household's ability to access finance and invest in productive activities. Investments in rural education and improving rural public health systems can help alleviate constraints that commonly afflict smallholder farming households (Schultz, 1998). Capacity will influence a household's willingness to pursue and ability to meet the requirements posed by modern chains. It will also affect a household's ability to access credit and reduce the burden of many transaction costs specific to modern chains (Barrett *et al.*, 2001). Capacity, in turn, is built through education and experience, among other factors.

Inclusion/exclusion happens at the farm-market interface. As long as there are entities or intermediaries that can buffer the scale-specific needs of buyers against the capabilities of the small-scale producer, and cover their costs by adding value, then there is no reason why smallholders should be excluded in a world where organized retail is expanding rapidly. However, because different strategies for bridging the scale mismatch are associated with different types of transaction costs, the appropriate model depends on the context.

Organization and cooperation seem to be natural responses to reducing transaction costs arising from the scale mismatch between individual farmers and those procuring from them. Local organizations are essential for the scaling up function, linking small scale producers with larger scale buyers. Without some local initiative, it is highly improbable that individual households can tap into modern chains. A critical threshold must be crossed, either by local producers who band together and pursue market opportunities, or by a buyer coming to a place with the purpose of procuring a product. As governments across the world are diminishing their institutional support for agriculture, local organizations are stepping in to fill the void. Organizations have tackled the challenges of marketing produce, adding value through processing, input provision, financial services and market information, and vocalizing key elements of the policy agenda. There has been no magic formula for developing these organizations or ensuring their effectiveness. But local capacity gives rise to leadership and transparency seems to promote perceptions of fairness, thus keeping members and clients satisfied (Shepherd, 2007).

In evaluating different strategies to link smallholders with markets, it is important to consider the cost of implementing them against the benefits, along with the distribution of costs and benefits between households. It would be hard to justify, by any reasonable cost-effective criteria, many supply chain development projects that have been carried out to promote smallholder participation in modern chains (Meijer *et al.*, 2008; Shepherd, 2007).

In general, interventions to facilitate smallholder participation in modern chains should not be heavy handed. A top priority is creating an enabling environment through the provision of public goods that reduce transaction costs. There are many ways to lower the costs of doing business with smallholders. Governments can leverage incentives for including smallholders without being directly involved. In India, the state government of Punjab provided incentives for contracting with farmers that included reimbursement of extension costs. Such incentives could, instead, be specifically targeted towards those who contract with smallholders, in order to negate the higher per-farmer transaction costs that the procuring company incurs by contracting with smallholders. Information asymmetry costs can be reduced with improved marketing extension and capacity building and market information systems. Improving market and transportation infrastructure, as well as transportation services, will reduce the transaction costs associated with negotiation as marketing costs are lowered and more marketing channels become available. Finally, improving the legal and institutional environment surrounding contract formulation and arbitration will reduce smallholders' costs of entering into more formal agreements by making them more available.

Public investments in specific chains and projects should be carefully considered. Picking 'winners' is problematic. Public investments should be weighed against the benefits they generate and how those benefits are distributed. Well-placed public chain investments can be catalytic. However, chain-specific investments to link smallholders into modern chains are likely to be costly in terms of the number of farmers reached and the income effect on each farmer. Such targeted investments benefit participants, but there are almost always few participants relative to non-participants, and there is a threat of further alienating non-participants. Governments will do best by supporting a competitive, investment-friendly environment that is also well regulated and allowing the market to pick the winners while leveraging maximum social benefit from private investments in modern chains.

4.3 Upgrading the traditional system

While benefits from investments in modern supply chains for specific sub-sectors are largely held by participants, investments in traditional wholesale markets are shared more equitably. Well functioning traditional markets facilitate procurement for modern retailers and processors who can avoid investing in alternate infrastructure to bypass the traditional system if it serves their needs. When retailer bypassing becomes widespread, incentives for upgrading traditional markets with public resources are reduced. Key advocates for upgrading the public system may be appeased if the private system meets their needs, leaving behind the traditional market and widening the gap between modern chains and traditional ones, leading to 'duality' in domestic food systems. Improvements in traditional markets also serve traditional processors and retailers, who do not have the option of bypassing traditional wholesale markets. Improvements even serve farmers who are participating in modern chains relative than traditional markets because modern buyers must compete with the traditional marketplaces.

Some simple improvements in market structure can improve traceability and public health standards while reducing spoilage rates. Improved flow of information and regulation of market transactions will reduce the transaction costs that arise from information asymmetry and trader corroboration. These interventions must be financed somehow, and they could have an impact on price margins. Traditional markets will never compete with vertically coordinated private chains in product differentiation and information exchange, but strategic investments in market structure could allow traditional markets to achieve a minimum standard that meets the needs of many retailers and processors.

In countries at the low end of the transformation process, the priority is in expanding traditional wholesale markets, improving their structure and forging upstream linkages with producers and downstream linkages with retailers and processors. In modernizing food systems, it is important to improve traceability and reduce spoilage rates in wholesale markets to reduce retailers' incentives for bypassing. The HACCP (Hazard Analysis and Critical Control Point) system offers some promise for implementing basic safety standards in traditional chains, but its implementation requires widespread education and cooperation throughout the chain (Unnevehr and Jensen, 1999). Opportunities for differentiating products based on quality attributes should be further explored within traditional wholesale markets, particularly in Asia where they still hold a large market share, so that traditional markets can continue to serve organized retailers, street vendors and everything in between. The wider the barriers separating traditional and modern chains, the more difficult and risky it is for a farmer to participate in modern chains relative to alternatives (Narrod *et al.*, 2008). When there is a healthy and domestic retail and/or processing sector, and a wholesale system that accounts for product differentiation, farmers have more options between the opposite extremes of basic traditional commodity markets and high value exports of fresh produce. In China, the vibrant traditional wholesale sector accommodated the full spectrum of quality needs.

Poor infrastructure for transport raises the price of inputs while lowering the price of outputs. Where infrastructure is poor, the input to output price ratio is a key determinant of competitiveness in a given location (Heisey and Mwangi, 1997). Post-harvest infrastructure for storage will improve marketing flexibility while decreasing the burden of spoilage. When wholesale market infrastructure and collection points are present, farmers have the opportunity to earn higher marketing margins. On the whole, investments in transport and marketing infrastructure will expand the range of consumers that farmers can reach while increasing the prices they can earn, lowering marketing risk while raising incomes.

4.4 Safety nets for non-participants

For many smallholders throughout the world, and particularly in sub-Saharan Africa, opportunities to participate in modern, organized chains are eclipsed by the more fundamental challenge of participating in any market. Market linkages for smallholders can be improved by lowering transaction costs, investing in market and transport infrastructure, boosting smallholder productivity and improving access to inputs. Supporting smallholder productivity is essential and benefits both non-sellers and sellers. When markets are thin and prices are variable, livestock and cassava can be harvested flexibly, allowing smallholders to manage price risk more effectively (Jayne, 2008). In countries where very few farmers are participating meaningfully in markets, commodity price supports benefit a few 'elite' farmers disproportionately (Jayne, 2008).

Many places are simply unsuitable for high-value agriculture because of agro-climatic limitations. Only a narrow range of agro-climatic conditions is suitable for rainfed horticulture, for instance. To an extent, physical factors determine the set of crops that can be grown in any place. However, local investments, for example in water resource development, can expand the set of options available, and the potential for diversification. Targeted investments in R&D for production in less-favoured areas can also help overcome agro-climatic constraints, but in most cases technical expertise must be brought in from other places. Production technology for less favoured areas may become disruptive in the long term, but a lack of foreseeable payoffs in the short term may deter sufficient investment.

Improved infrastructure can possibly alleviate land constraints. In Zambia, for example, land holdings are clustered in higher potential areas, with lower potential and more remote areas being less inhabited (Jayne, 2008). Improving road infrastructure could effectively increase a country's productive land area. In some instances, smallholders in isolated areas and low input production systems may be competitive in local markets because poor infrastructure raises retail prices. While improving transport infrastructure may reduce some farmers' ability to market some crops profitability, the benefits of infrastructure expansion are likely to be shared more widely and outweigh the costs. In southern and eastern Africa, there are many more buyers of staple grains than sellers (Jayne, 2008). As needed, mechanisms can be devised to compensate those who are hurt by infrastructure expansion.

Even after improving productivity and market access, many smallholders will remain in production on a subsistence basis or will pursue off-farm income or migrate to towns and cities. Through the process of agricultural transformation, it is normal for the size of the population dependent on agriculture to decrease over time as agriculture's share in the economy falls and as per capita incomes rise. When migration out of rural areas occurs faster than growth in opportunities to earn income in rural areas, this migration results in a transfer of poverty rather than true poverty reduction associated with the agricultural transformation (Ravallion *et al.*, 2007). Developing alternative incomes in rural areas is essential for seeing smallholders who have no future in farming through the transition. Social safety nets, such as targeted feeding programmes for chronic hunger victims, are essential for those who have no sources of income and limited prospects.

Discussion in lecture room

Apply your theoretical knowledge with respect to this paper:
- What are the barriers for smallholders to participate in the transition of food systems? Do you see any differences in how the conventional food system is dealing with smallholders

compared to how the organic food system is dealing with small organic farmers? Also see Kasterine *et al.* (2010) about contract farming and certification costs.
- The Royal Swedish Academy of Sciences compiled a document with the title *Trade and geography – economies of scale, differentiated products and transport costs* (see http://nobelprize. org/nobel_prizes/economics/laureates/2008/ecoadv08.pdf). Read this paper and discuss the economic findings on migration and urbanization based on the theory of economies of scales. Which policy measures should governments in developing countries pursue to promote rural development and reduce rural poverty?

References

Anriquez, G. and Stamoulis, K., 2007. Rural development and poverty reduction: is agriculture still the key? ESA Working Paper no 07-02, Agricultural Development Economics Division, FAO, Rome, Italy.

Ashley, C. and Maxwell, S., 2001. Rethinking rural development. Development Policy Review 19: 395-425.

Balsevich, F., Berdegue, J., Flores, L., Mainville, D. and Reardon, T., 2003. Supermarkets and produce quality and safety standards in Latin America. American Journal of Agricultural Economics 85: 1147-1154.

Barboza, D., 2007. 774 arrests in China over safety. International Herald Tribune, 29 October.

Bardhan, P.K., 1983. Labor-tying in a poor agrarian economy: A theoretical and empirical analysis. The Quarterly Journal of Economics 98: 501-514.

Barret, C., Reardon, T. and Webb, P., 2001. Non farm income diversification and household livelihood strategies in rural Africa: Concepts, dynamics and policy implications. Food Policy 26: 315-331.

Barry, P.J., Sonka, S.T. and Lajili, K., 1992. Vertical coordination, financial structure, and the changing theory of the firm. American Journal of Agricultural Economics 74: 1219-1225.

Bénassy-Quéré, A., Coupet, M. and Mayer, T., 2007. Institutional determinants of Foreign Direct Investment. The World Economy 30: 764-782.

Beng-Huat, C., 2000. Consumption in Asia: Lifestyles and Identities. Routledge, London, UK.

Berdegue, J., Balsevich, F., Flores, L. and Reardon, T., 2005. Central American supermarkets private standards of quality and safety in procurement of fresh fruits and vegetables. Food Policy 30: 254-269.

Berry, R. and Cline, W., 1979. Agrarian Structure and Productivity in Developing Countries. John Hopkins University Press, Baltimore, MD, USA.

Bijman, W.J. and Hendrikse, G.W., 2003. Cooperatives in chains: Institutional restructuring in the Dutch fruit and vegetables industry. ERIM Report Series Research in Management, no 089-ORG.

Bingen, J., Serrano, A. and Howard, J., 2003. Linking farmers to markets: Different approaches to human capital development. Food Policy 28: 405-419.

Boselie, D., Henson, S. and Weatherspoon, D., 2003. Supermarket procurement practices in developing countries: Redefining the roles of the public and private sectors. American Journal of Agricultural Economics 85: 1155-1161.

Caldhilon, J.J., Moustier, P., Poole, N.D., Giac Tam, P.T. and Fearne, A.P., 2006. Traditional vs. modern food systems: Insights from vegetable supply chains to Ho Chi Minh City (Vietnam). Development Policy Review 24: 31-49.

Cassman, K., 1999. Ecological intensification of cereal production systems: Yield potential, soil quality and precision agriculture. Proceedings of the National Academy of Sciences USA 96: 5952-5959.

Chen, K., 2004. Retail revolution, entry barriers and emerging agri-food supply chains in selected Asian countries: Determinants, issues and policy choices. FAO/AGS, Rome, Italy.

Chen, K., Shepherd, A. and da Silva, C., 2005. Changes in food retailing in Asia: implications of supermarket practices for farmers and traditional marketing systems. Agricultural Management, Marketing and Finance Occasional Paper 8, FAO/AGS, Rome, Italy.

Chen, K. and Stamoulis, K., 2008. The changing nature and structure of agri-food systems in developing countries: Beyond the farm gate. In: E.B. McCullough, Pr. L. Pingali, and K.G. Stamoulis (eds.), The Tranformation of Agri-Food. Impacts on Smallholder Agriculture. FAO and Earthscan, London, pp. 143-157.

Coe, N. and Hess, M., 2005. The internationalization of retailing: implications for suppy network restructuring in east Asia and eastern Europe. Journal of Economic Geography 5: 449-473.

Crawford, E., Kelly, V., Jayne, T.S. and Howard, J., 2003. Input use and market development in Sub-Saharan Africa. Food Policy 28: 277-292.

D'Andrea, G., Lopez-Aleman, B. and Stangel, A., 2006. The supermarket's revolution in developing countries: Policies to address emerging tensions among supermarkets, suppliers and traditional retailers. The European Journal of Development Research 18: 522-545.

Davis, B., Winters, P., Carletto, G., Covarrubias, K., Quinones, E., Zezza, A., Stamoulis, K., Bonomi, G. and DiGiusseppe, S., 2007. Rural income generating activities: A cross country comparison. ESA Working Paper No. 07-16, FAO, Rome, Italy.

Dawe, D., 2005. Economic growth and small farms in Suphan Buri, Thailand. Paper prepared for the symposium, Agricultural Commercialization and the Small Farmer, Agricultural and Development Economics Division (ESA), FAO, Rome, Italy.

Digal, L., 2004. Quality grading in the supply chain: The case of vegetables in southern Philippines. Journal of International Food and Agribusiness Marketing 17: 71-93.

Dirven, M. and Faiguenbaum, S., 2008. The role of Santiago wholesale markets in supporting small farmers and poor consumers. In: E.B. McCullough, Pr. L. Pingali and K.G. Stamoulis (eds.), The Tranformation of Agri-Food. Impacts on Smallholder Agriculture. FAO and Earthscan, London, UK, pp. 171-188.

Dolan, C. and Humphery, J., 2002. Changing governance patterns in trade in fresh vegetables between Africa and the United Kingdom. www.gapresearch.org/production/IFAMSSubmission.pdf. Accessed January 2007.

Dries, L., 2005. The impact of supermarket development on the rural economy. Conceptual framework draft 21 July.

Dries, L., Reardon, T. and Swinnen, J., 2004. The rapid rise of supermarkets in central and eastern Europe: Implications for the agrifood sector and rural development. Development Policy Review 22: 525-556.

Dries, L. and Swinnen, J., 2008. The impact of globalization and vertical integration in agri-food processing on local suppliers: Evidence from the Polish dairy sector. In: E.B. McCullough, Pr. L. Pingali and K.G. Stamoulis (eds.), The Tranformation of Agri-Food. Impacts on Smallholder Agriculture. FAO and Earthscan, London, UK.

Eswaran and Kotwal, 1985. A theory of contractual structure in agriculture. The American Economic Review 75: 352-267.

Falcon, W., Timmer, C. and Pearson, S., 1983. Food Policy Analysis. John Hopkins University Press, Baltimore, MD, USA.

Fan, S. and Chan-Kang, C., 2004. Returns to investment in less favoured areas in developing countries: A synthesis of evidence and implications for Africa. Food Policy 29: 431-444.

FAO, 2004. The state of food insecurity in the world. FAO, Rome, Italy.

FAO, 2006. World Agriculture: Towards 2030/2050, Interim Report, FAO, Rome, Italy.

Filmer, D. and Pritchett, L., 1999. The effect of household wealth on education attainment: Evidence from 35 countries. Population and Development Review 25: 85-120.

Foer, A., 2001. Small business and antitrust. Small Business Economics 16: 3-20.

Fransoo, J. and Wouters, M., 2000. Measuring the bull whip effect in the supply chain. Supply Chain Management 5: 78-89.

Gibbon, P., 2003. Value-chain governance, public regulation, and entry barriers in the global fresh fruit and vegetable chain into the EU. Development Policy Review 21: 615-625.

Globerman, S. and Shapiro, D., 2002. Global Foreign Direct Investment flows: The role of governance infrastructure. World Development 30: 1899-1919.

Golan, E., Krissoff, B., Kuchler, F., Calvin, L., Nelson, K. and Price, G., 2004. Traceability in the U.S. food supply: Economic theory and industry studies. Agricultural Economic Report No 830, Economic Research Service, US Department of Agriculture, Washington, DC, USA.

Gow, H. and Swinnen, J., 1998. Up and downstream restructuring, Foreign Direct Investment and hold-up problems in agricultural transition. European Review of Agricultural Economics 25: 331-350.

Hallam, D., Liu, P., Lavers, G., Pilkauskas, P., Rapsomanikis, G. and Claro, J., 2004. The market for non-traditional agricultural exports. FAO Commodities and Trade Technical Paper, FAO, Rome, Italy.

Heisey, P. and Mwangi, W., 1997. Fertilizer use and maize production. In: D. Byerlee and C. Eicher (eds.), Africa's Emerging Maize Revolution, Lynne Rienner Publishers, Boulder, CO, USA, pp. 193-211.

Helfand, S. and Levine, E., 2004. Farm size and the determinants of productive efficiency in the Brazilian Center-West. Agricultural Economics 31: 241-249.

Heltberg, R., 1998. Rural market imperfections and the farm size – productivity relationship evidence from Pakistan. World Development 26: 1807-1826.

Hu, D., Reardon, T., Rozelle, S., Timmer, C. and Wang, H., 2004. The emergence of supermarkets with Chinese characteristics: Challenges and opportunities for China's agricultural development. Development Policy Review: 557-586.

Huang, J., Wu, Y. and Rozelle, S., 2008. Marketing China's Fruit: Are small, poor farmers being excluded from the supply chain? In: E.B. McCullough, Pr., L. Pingali and K.G. Stamoulis (eds.), The Tranformation of Agri-Food. Impacts on Smallholder Agriculture. FAO and Earthscan, London, UK, pp. 311-331.

Jayne, T.S., 2008. Forces of change affecting African food markets: Implications for public policy. In: : E.B. McCullough, Pr.,L. Pingali and K.G. Stamoulis (eds.), The Tranformation of Agri-Food. Impacts on Smallholder Agriculture. FAO and Earthscan, London, pp. 109-140.

Kennedy, E. and Reardon, T., 1994. Shift to non-traditional grains in the diets of east and west Africa: Role of women's opportunity cost of time in prepared-food consumption. Food Policy 19: 45-56.

Kimle, K. and Hayenga, M., 1993. Structural change among agricultural input industries. Agribusiness 9: 15-27.

King, R. and Phimpiu, P., 1996. Reengineering the food supply chain: The ECR initiative in the grocery industry. American Journal of Agricultural Economics 78: 1181-1186.

King, R. and Venturini, L., 2005. Demand for quality drives changes in food supply chains. New directions in global food markets, A1b–794. Economic Research Service USDA.

Kinsey, J. and Senauer, B., 1996. Consumer trends and changing food retailing formats. American Journal of Agricultural Economics 78: 1187-1191.

Kirsten, J. and Sartorius, K., 2002. Linking agribusiness and smale-scale farmers in developing countries: Is there a new role for contract farming?. Development Southern Africa 19: 503-529.

Kumar, K., 2001. Technology for supporting supply. Communications of the Association for Computing Machinery (ACM), vol 44, no 6.

Kumar, P., 2005. Commercialization of Indian agriculture and its implications for small and large farmers: A case study of Punjab. Paper prepared for the symposium, Agricultural Commercialization and the Small Farmer, Agricultural and Development Economics Division (ESA), FAO, Rome, Italy.

Lundy, M., Banegas, R., Centeno, L., Rodriguez, I., Alfaro, M., Hernandez, S. and Cruz, J.A., 2006. Assessing small-holder participation in value chains: The case of vegetables in Honduras and El Salvador. FAO, Rome, Italy.

Maertens, M. and Swinnen, J.F.M., 2006. Trade, standards and poverty: Evidence from Senegal. Licos discussion paper no177/2006, Catholic University of Leuven.

Maruyama, M. and Viet Trung, L., 2007. Supermarkets in Vietnam: Opportunities and obstacles. Asian Economic Journal 21: 19-46.

McCluskey, J., 2007. Public and private food quality standards: Recent trends and strategic incentives. In: J. Swinnen (ed.), Global Supply Chains, Standards and the Poor. CABI International, Wallingford, UK, pp. 19-25.

McCullough, E. and Pingali, P.L., 2008. Overview of case studies assessing impacts of food systems transformation on smallholder farmers. In: E.B. McCullough, Pr., L. Pingali and K.G. Stamoulis (eds.), The Tranformation of Agri-Food. Impacts on Smallholder Agriculture. FAO and Earthscan, London, pp. 227-234.

McCullough, E.B., Pingali, Pr., L. and Stamoulis, K.G. (Eds.), 2008. The transformation of agrifood systems: globalization, supply chains and smallholder farmers. FAO and Earthscan, London, UK.

McKean, J. D., 2001. The importance of traceability for public health and consumer protection. Revue Scientifique el Technique de l'Office International Epizooties 20: 363-371.

Meijer, M., Rodriguez, I., Lundy, M. and Hellin, J., 2008. Supermarkets and small farmers: The case of fresh vegetables in Honduras. In: E.B. McCullough, Pr., L. Pingali and K.G. Stamoulis (eds.), The Tranformation of Agri-Food. Impacts on Smallholder Agriculture. FAO and Earthscan, London, pp. 333-353.

Morris, M.L., Rusike, J. and Smale, M., 1998. Maize seed industries: A conceptual framework. In: M.L. Morris (ed.), Maize Seed Industries in Developing Countries. Lynne Rienner Publishers, Boulder, CO, USA, pp. 35-54.

Narrod, C., Roy, D., Avendaño, B. and Okello, J., 2008. Impact of international food safety standards on smallholders: Evidence from three cases. In: E.B. McCullough, Pr., L. Pingali and K.G. Stamoulis (eds.), The Tranformation of Agri-Food. Impacts on Smallholder Agriculture. FAO and Earthscan, London, pp. 355-372.

Neven, D. and Reardon, T., 2008. The rapid rise of Kenyan supermarkets: Impacts on the fruits and vegetables supply system. In: E.B. McCullough, Pr., L. Pingali and K.G. Stamoulis (eds.), The Tranformation of Agri-Food. Impacts on Smallholder Agriculture. FAO and Earthscan, London, pp. 189-206.

OECD, 2007. Stat database, available at: www.oecd.org/statsportal. Accessed January 2007.

Omiti, J., Nyanamba, T., McCullough, E. and Otieno, D., 2008. The transition from maize production systems to high-value agriculture in Kenya. In: E.B. McCullough, Pr., L. Pingali and K.G. Stamoulis (eds.), The Tranformation of Agri-Food. Impacts on Smallholder Agriculture. FAO and Earthscan, London, pp. 235-257.

Opara, L., 2003. Traceability in agriculture and food supply chain: A review of basic concepts, technological implications and future prospects. Food Agriculture and Environment 1: 101-106.

Pingali, P., 1997. From subsistence to commercial production systems: The transformation of Asian agriculture. American Journal of Agricultural Economics 79: 628-634.

Pingali, P., 2006. Agricultural growth and economic development: A view through the globalisation lens. Presidential Address to the 26th International Conference of Agricultural Economists, 12-18 August, Gold Coast, Australia.

Pingali, P., 2007. Westernization of Asian diets and the transformation of food systems: Implications for research and policy. Food Policy 32: 281-298.

Pingali, P., Khwaja, Y. and Meijer, M., 2007. The role of the public and private sectors in commercializing small farms and reducing transaction costs. In J.F.M Swinnen (ed.), Global Supply Chains, Standards and the Poor, CABI International, Wallingford, UK, pp. 267-280.

Popkin, B., 1999. Urbanization, lifestyle changes and the nutrition transition. World Development 27: 1905-1916.

Population Division of the Department of Economic and Social Affairs of the United Nations Secretariat, 2006. World Population Prospects: The 2006 Revision and World Urbanization Prospects: The 2005 Revision.

Rademakers, M., 2000. Agents of trust: Business associations in agri-food supply systems. International Food and Agribusiness Management Review 3: 139-153.

Ravallion, M., Chen, S. and Sangraula, P., 2007. New evidence on the urbanization of global poverty. Background Paper for the WDR 2008.

Reardon, T. and Berdegue, J., 2002. The rapid rise of supermarkets in Latin America: Challenges and opportunities for development. Development Policy Review 20: 371-388.

Reardon, T. and Hopkins, R., 2006. The supermarket revolution in developing countries: Policies to address emerging tensions among supermarkets, suppliers and traditional retailers. The European Journal of Development Research 18: 522-545.

Reardon, T., Stamoulis, K. and Pingali, P., 2007. Rural nonfarm employment in developing countries in an era of globalization. In: K. Otsuka and K. Kalirajan (eds.), Contributions of Agricultural Economics to Critical Policy Issues: Proceedings of the Twenty-Sixth Conference of the International Association of Agricultural Economists, Brisbane, Australia, 12–18 August, Blackwell Publishing, Malden.

Reardon, T., Timmer, C. and Berdegue, J., 2008. The rapid rise of supermarkets in developing countries: Induced organizational, institutional and technological change in agri-food systems. In: E.B. McCullough, Pr., L. Pingali and K.G. Stamoulis (eds.), The Tranformation of Agri-Food. Impacts on Smallholder Agriculture. FAO and Earthscan, London, pp. 47-66.

Regmi, A. and Dyck, J., 2001. Effects of urbanization on global food demand. In: A. Regmi (ed.), Changing Structures of Global Food Consumption and Trade, Economic Research Service, United States Department of Agriculture, Washington DC, USA, pp. 23-30.

Rosegrant, M., Cai, X., Cline, S. and Nakagawa, N., 2002. The role of rainfed agriculture in the future of global food production. EPTD discussion paper no 90, International Food Policy Research Institute, Washington, DC, USA.

Sadoulet, E. and De Janvry, A., 1995. Quantitative Development Policy Analysis. Johns Hopkins University Press, Baltimore, MD, USA.

Schultz, T., 1988. Education investments and returns. In: H. Chenery and T.N. Srinivasan (eds.), Handbook of Development Economics, Volume 1, Elsevier Science Publishers, Amsterdam, pp. 543-630.

Schwentesium, R. and Gomez, M.A., 2002. Supermarkets in Mexico: Impacts on horticulture systems. Development Policy Review 20: 487-502.

Shepherd, A., 2007. Approaches to linking producers to markets. Agricultural Management, Marketing and Finance Occasional Paper 13, Rural Infrastructure and Agro-Industries Division, FAO, Rome, Italy.

Simmons, P., Winters, P. and Patrick, I., 2005. An analysis of contract farming in east Java, Bali and Lombok, Indonesia. Agricultural Economics 33: 513-525.

Singh, I., Squire, L. and Strauss, J., 1986. Agricultural Household Models: Extensions, Applications and Policy. Johns Hopkins University Press, Baltimore, MD, USA.

Singh, S., 2002. Contracting out solutions: Political economy of contract farming in the Indian Punjab. World Development 30: 1621-1638.

Singh, S., 2008. Marketing channels and their implications for smallholder farmers in India. In: E.B. McCullough, Pr., L. Pingali and K.G. Stamoulis (eds.), The Tranformation of Agri-Food. Impacts on Smallholder Agriculture. FAO and Earthscan, London, pp. 279-310.

Solamino, A., 2006. Asset accumulation by the middle class and the poor in Latin America: political economy and governance dimensions. Macroeconomics of Development, no 55, Economic Commision for Latin American and the Caribbean (ECLAC).

Steinfeld, H. and Chilonda, P., 2005. Old players, new players: Livestock report 2006. FAO, Rome, Italy.

Swinnen, J., 2002. Transaction and integration in Europe: Implications for agricultural and food markets, policy and trade agreements. The World Economy 25: 481-501.

Swinnen, J., 2005. Small farms, transition and globalisation in Central and EasternEurope and the former Soviet Union. Paper prepared for the symposium, Agricultural Commercialization and the Small Farmer, Agricultural and Development Economics Division (ESA), FAO, Rome, Italy.

Swinnen, J., 2007. Global Supply Chains, Standards and the Poor: How Globalization of Food Systems and Standards Affects Rural Development and Poverty, CABI International, Wallingford, UK.

Timmer, C. P., 2008. Food policy in the era of supermarkets: What's different? In: E.B. McCullough, Pr., L. Pingali and K.G. Stamoulis (eds.), The Tranformation of Agri-Food. Impacts on Smallholder Agriculture. FAO and Earthscan, London, pp. 67-86.

Tobgay, S. and McCullough, E.B., 2008. Linking small farmers in Bhutan with markets: The importance of road access. In: E.B. McCullough, Pr., L. Pingali and K.G. Stamoulis (eds.), The Tranformation of Agri-Food. Impacts on Smallholder Agriculture. FAO and Earthscan, London, pp. 259-278.

Traill, W., 2006. The rapid rise of supermarkets? Development Policy Review 24: 163-174.

Unnevehr, L. and Jensen, H., 1999. The economic implications of using HACCP as a food safety regulatory standard. Food Policy 24: 625-635.

Unnevehr, L. and Roberts, T., 2002. Food safety incentives in a changing world food system. Food Control 13: 73-76.

Vorley, B., 2003. Food, Inc. Corporate concentration from farm to consumer. UK Food Group, London, UK.

Weatherspoon, D. and Reardon, T., 2003. The rise of supermarkets in Africa: Implications for agrifood systems and the rural poor. Development Policy Review 21: 333-355.

Welch, R. and Mitchell, P., 2000. Food processing: a century of change. British Medical Bulletin 56: 1-17.

Wilkinson, J., 2008. The food processing industry, globalization and developing countries. In: E.B. McCullough, Pr., L. Pingali and K.G. Stamoulis (eds.), The Tranformation of Agri-Food. Impacts on Smallholder Agriculture. FAO and Earthscan, London, pp. 87-108.

World Bank, 2006. World Development Indicators, World Bank, Washington DC, USA.

World Bank, 2008. World Development Report 2008: Agriculture for Development, World Bank, Washington, DC, USA.

Role of certification bodies in the organic production system

M. Canavari[1], N. Cantore[2,3], E. Pignatti[1] and R. Spadoni[1]

[1]Alma Mater Studiorum University of Bologna (UniBO), Department of Agricultural Economics and Engineering, Viale Fanin 50, 40127 Bologna, Italy
[2]Overseas Development Institute, 111 Westminster Bridge Road, London SE1 7JD, United Kingdom
[3]Università Cattolica del Sacro Cuore, Largo Agostino Gemelli 1, 20123 Milan, Italy

Abstract

In this paper the activity of certification bodies in the organic supply chain is analysed in a broad perspective. A general description of organic certification in different regions is provided, with a brief discussion of the harmonization and mutual recognition issues stressing how differences in regulations and laws enforcement generate transaction costs. We investigate the performance of third-party certification bodies in Italy and provide preliminary findings about their objectiveness and independence. Certification bodies play a crucial role in the organic supply chain, guaranteeing a correct information flow from producers to consumers through a quite complex environment. We evaluate the performance of third-party certification bodies by collecting information about measurable performance variables representing the intensity and quality of the certification bodies effort, with the aim to identify opportunistic organic producers' behaviour. Using a set of indicators for quality, intensity and frequency of inspections we analysed similarities and differences among certification bodies and we found a great heterogeneity in their operations management. This is very useful to describe different types of certification bodies. The drawback of our analysis is that we cannot truly identify the nature of these differences and we were not able to discriminate between virtuous and opportunistic certification bodies.

1. Introduction

Organic food products are an example of quality food products, i.e., products possessing a particular feature (attribute) that distinguishes them from other products of the same type and may make these products preferred on the market. When talking about quality and product attributes, economists distinguish between 'search goods' for which the quality attributes are apparent and can be verified before purchase, 'experience goods' for which the quality attributes are difficult to observe in advance, but they can be ascertained upon consumption and 'credence goods' for which consumers cannot easily ascertain the quality attributes even after they have consumed/used the goods (Darby and Karni, 1973; Nelson, 1970). Markets are able to easily manage quality differences when they are defined by search attributes, since people may simply choose the product they prefer and/or pay a higher price for it. Experience attributes are more difficult to manage, but mechanisms linked to product/producer/distributor reputation help to solve the problem. With credence goods sellers know the actual quality of the good while buyers do not, and this may give rise to market inefficiencies and fraud.

With credence goods the market may not be able to optimally allocate the resources (a situation that is usually addressed as 'market failure') because of asymmetric information and moral hazard. Asymmetric information is the situation in which one side of an economic relationship has more information than the other. Moral hazard is defined as the behaviour of agents who do not bear the full cost of their actions and thus are more likely to take such actions (OECD glossary: available at http://stats.oecd.org/glossary/). In other words agents are characterized by opportunistic behaviours to enjoy an advantage when they can skip the negative consequences.

In this paper, we discuss the nature of credence goods of organic food and we illustrate the mechanisms underlying the need to establish certification systems in order to provide quality signalling tools in this kind of market. In the following sections, we offer a double perspective of the issues concerning the organic certification systems. First, we briefly discuss the development of international standards for organic agriculture in different world areas and the ways by which they affect international trade of organic food. Second, we discuss the case of Italy, where we investigate the performance of certification bodies in a domestic market perspective.

2. Organic food as a credence good

Organic food (i.e., the 'organic' attribute for an agricultural produce, raw material, food product) may be classified in the category 'credence goods', since most of its features are inherent to the production process and cannot actually be detected in the final product.

The adoption of moral hazard behaviours is facilitated because the inspection costs to verify attributes for these goods are high (Darby and Karni, 1973). For organic food moral hazard is induced by producers who are interested in receiving the premium price for quality attributes they claim to possess, but without adopting the (generally more expensive) production procedures that are necessary to ensure the required quality attribute.

Therefore, the main problem faced by consumers who try to maximize their utility is asymmetric information, since they are not fully aware of the choice made by agents in the upper side of the agri-food supply chain. This information gap for consumers is further increased by difficulties in the definition of food quality attributes.

In the context of organic food marketing, the role of control measures in monitoring food quality is crucial (McCluskey, 2000). This monitoring role can be assumed by public authorities for quality attributes that are considered as non-tradable public goods and must be taken for granted in order to make the market work properly, such as food safety attributes. In other cases, quality attributes may be requested by certain markets or segments, therefore they may become tradable.

In both cases, however, the private players must assume the responsibility to assure the quality level through proper operations management, monitoring and control and by implementing an effective quality assurance system. Therefore, quality is basically assured by the producer itself and may be communicated to the market through a quality signal represented by the producer's brand; in addition, a customer (for instance, the retailer towards the producer) may play a role in assuring the quality of the product, both controlling the product and monitoring the producer operations and its quality assurance system through on-site inspections; finally, this control and monitoring function can be effectively played by third-party certification bodies that can guarantee transparent and independent procedures, giving assurance to the interested parties that controls and monitoring are properly implemented within the producer's quality assurance system.

Third-party certifiers are committed to provide certifications as quality signals to equilibrate the information gap existing between producers and consumers. Certifications and brands both work using a reputation mechanism and provide the market with the necessary information about credence goods such as quality food that consumers had difficulty acquiring. To a certain extent, their purpose is to transform credence goods into search goods, for which the quality is easily recognized using the certification seal or the brand logo. However, brands and certification seals may be in competition and play different roles, since the former relies upon the company/

product's reputation and it only belongs to its owner, while the latter relies upon the standard/ certification body's reputation and can be used by many competing companies.

In other words, certification guarantees production processes and quality goods features and may represent an important market tool for products differentiation, but it cannot substitute branding. The mechanism underlying the elimination of asymmetric information distortions through the provision of reliable certifications to signal product quality and attributes can be affected by moral hazard behaviours of third-party certification bodies. Perfect competition among certification bodies would reduce inspection costs, but would affect the reliability of third-party certifiers that could reduce the intensity of controls to attract new customers, especially in the market segment of those producers who practice opportunistic behaviours. This phenomenon would create a risky vicious circle. Moral hazard of third-party certifiers could induce and encourage moral hazard of producers (Giannakas, 2002). Opportunistic behaviours of the supply chain agents could generate a risky mechanism of adverse selection.

Adverse selection is a mechanism by which if the signalling mechanism does not work, consumers cannot distinguish between high quality and regular quality goods and their willingness to pay for quality goods is lower. Non-opportunistic producers are slowly excluded by the market because consumers cannot acknowledge the quality of their products. As outlined by Jahn *et al.* (2005), reputation mechanisms, the improvement of audit procedures and the adoption of rotation procedures could reduce the risk of moral hazard behaviours of certification bodies.

Certification mechanisms should also fit the market features of specific supply chains. For the organic sector, it is necessary that the certification systems take into account the specific issues that may affect the level of confidence of market operators and consumers that the organic products available on the market actually meet their needs and expectations. The slow growth of the organic market in many countries all around the world, the strong debate about the definition of what is organic and a certain scepticism about organic food suggest that particular attention should be paid to certification in the organic food sector and that analysis and careful design of the third-party certification system is needed.

3. Organic certification systems around the world: consequences for international trading

Certification is defined as 'The procedure by which a third-party (certification body) gives written assurance that a clearly identified process has been methodically assessed in a way that provides adequate confidence that specified products conform to specified standards' (ISO/IEC 17000/2004). Those specific requirements may involve products, processes, systems or persons. The certification may give rise to a third-party attestation, that is the issue of a statement based on a decision following review that fulfilment of specified requirements has been demonstrated.

Certification should be performed by accredited bodies, where accreditation is the procedure by which an authoritative body or accreditor gives a formal recognition that a certification body is competent to carry out certification according to specific standards. Examples of 'product oriented' certification standards are 'fair trade', which is useful to identify those products matching international solidarity and equity requirements, British Retail Consortium, for successful and responsible retailing (BRC), International Food Standard (IFS) and Global Partnership for Good Agricultural Practice (GlobalGAP). The focus is mainly on aspects related to safety, hygienic and sanitary conditions. Among 'systems oriented' certification standards we mention ISO22000:2005 (Requirements for food safety management systems), ISO22005:2007 (Traceability in the feed and food chain), ISO9001:2008 (Quality management systems) and ISO14001:2004 (Environmental management systems).

Another useful distinction about certification is that it can be *mandatory* or *voluntary*. In the first case the national government sets up standards and defines the rules for their implementation by law in order to preserve health or to promote targets of public interest. Voluntary certification is mainly a tool that firms can choose in order to differentiate their products. Compliance to mandatory standards, instead, does not allow any differentiation since it is necessary to run the business.

As predicted by the basic microeconomic theory in a perfect competition set up, firms supplying homogenous products cannot impose high prices and increase profits, therefore differentiation is a means by which they can act as 'monopolists' in offering for a specific product one or more different attributes matching particular consumers' tastes. If a firm is making a profit selling a product in an industry, and other industries are not allowed to perfectly reproduce that product, they still may find it profitable to enter that industry and produce a similar but distinctive product. Economists refer to this phenomenon as product differentiation (Varian, 1987). All the certification standards we previously mentioned ('product oriented' and 'system oriented') are not mandatory but they represent opportunities for agri-food firms to distinguish their product in the agri-food market. The standards do not only represent a way to improve quality and to differentiate the company in the business arena, they usually also provide internal benefits in terms of a better organisation and of improved internal and customer relationships management. The compliance with the organic farming standards, however, is usually mandatory for companies claiming to produce organic products in countries where organic agriculture and organic food production have been regulated. Therefore, once a company/farm decides to convert to organic farming and to sell organic products, they have to comply with the mandatory standard.

A wide set of definitions tries to describe organic farming. One of the most popular is represented by the United States Department of Agriculture (USDA) definition (1995): 'An ecological production management system that promotes and enhances biodiversity, biological cycles, and soil biological activity. It is based on minimal use of off-farm inputs and on management practices that restore, maintain, or enhance ecological harmony. The primary goal of organic agriculture is to optimize the health and productivity of interdependent communities of soil life, plants, animals and people'.

A general definition usually provides a basic guideline to address organic farming activities by drawing the main principles. The USDA definition attempts to underline the role of organic farming as positive externality for the society benefiting the environment and the ecological equilibrium. The mission of every policy maker is to transform a definition into concrete measures to yield consistent product characteristics and procedures. At the international level we observe different organic certification systems for many countries as laws generally reflect different cultural, political and social conditions. Different standards and procedures generate market frictions and transaction costs as specific recognised third-party bodies are charged to verify the correspondence of foreign organic products to national organic standards. This inspection activity is of course reflected in higher business costs. In other words, a lack of trust in international trading relationships at international level raises the costs of undertaking bilateral commercial relationships. In spite of this point of view different standards and procedures may increase the volume of world trading. If consumers' preferences were oriented to accept only specific national standards, a process of convergence at international level may not meet market needs and decrease business exchanges. However, evidence shows that this effect is not relevant and that standards differentiation is mainly trade-reducing (Cranfield *et al.,* 2009).

Organic certification is included among those systems that are regulated by law in many countries. Basically, when organic agriculture is regulated by the national government, firms can freely choose to adopt organic practices in order to differentiate their product, but when they choose to produce organic food they have to follow rules determined by the specific regulation.

In the global agri-food market the most important certification systems are represented by the ones in force in the European Union, in the United States (NOP), and in Japan (JAS). Where a national regulation does not exist or is only partially implemented, organic certification may follow foreign rules (such as EU standards) or the standards set up by international organizations (such as the International Federation of Organic Agriculture Movements, IFOAM).

As a consequence of this situation, a certification body may be accredited by several accreditors, thus being able to issue different types of organic certification. An International Task Force on Harmonization and Equivalence in Organic Agriculture (ITF) has developed a document defining the International Requirements for Organic Certification Bodies (IROCB), which is based upon the requirements in ISO/IEC Guide 65: 1996 (E) 'General requirements for bodies operating product certification systems.' However, given that organic certification has special features that differ from certification of products and services covered by ISO/IEC Guide 65, IROCB also takes into account the IFOAM Accreditation Criteria for Bodies Certifying Organic Production and Processing (IAC) and includes sector-specific requirements. It also includes reformulated and amended International Organization for Standardization (ISO) paragraphs and additional requirements to cover issues confronting a certification body when undertaking organic certification (UNCTAD/FAO/IFOAM, 2008)

In the member states of the European Union (EU), the labelling of plant products as organic has been governed by an EU Regulation that came into force in 1993, while products from organically managed livestock are governed by an EU Regulation enacted in August 2000. The EU Regulation on organic production sets up rules governing the production, processing and import of organic products, including inspection procedures, labelling and marketing, for the whole of Europe. Each European country has been responsible for enforcement and for its own monitoring and inspection system. Applications, supervision and sanctions are dealt with at regional levels. At the same time, each country has the responsibility to interpret the regulation on organic production and to implement the regulation in its national context. Altogether, more than two hundreds certification bodies are currently accredited in the EU and the certification and inspection framework is different across the EU member states.

The regulations currently in force in the EU (Reg. EC 834/07, 889/2008 and 1235/08) introduce important rules for organic farming. The use of the EU label will become compulsory from the 1st of July 2010 on, but it can be introduced together with other private labels. A product cannot be defined as organic if it contains a level of organic ingredients that is lower than 95% or a level of Genetically Modified (GM) ingredients higher than 0.9% for each ingredient. The import procedures from non European countries are different if specific countries enjoy privileged import procedures through equivalence agreements. Simplified procedures are granted to those operators importing organic products from Argentina, Australia, Israel, India, Switzerland, Costa Rica and New Zealand.

In the United States, the National Organic Program (NOP) is the federal regulatory framework governing organic food. The regulation was issued in October 2002 and is administered by USDA. The Organic Food Production Act of 1990 required that the USDA develop national standards for organic products. The regulations are enforced by the USDA through the NOP under this act.

It covers in detail all aspects of food production, processing, delivery and retail sale. Under the NOP, farmers and food processors who wish to use the word 'organic' in reference to their businesses and products must be certified organic. Producers with annual sales not exceeding 5,000 US$ are exempted and do not require certification. A USDA Organic seal identifies products with at least 95% organic ingredients.

There are currently 56 US domestic certification agencies accredited by the USDA, including Organic Crop Improvement Association, CCOF, Quality Assurance International (QAI), and Indiana Certified Organic. There are also 41 accredited foreign agencies that offer organic certification services. The NOP covers fresh and processed agricultural food products, including crops and livestock. US accepted few accreditation procedures of foreign governments. Certification bodies accredited according to the US requirements by Denmark, UK, India, Israel, New Zealand and Quebec are accepted by the USDA for certifying according to the US National Organic Program (NOP) without being directly accredited by United States Department of Agriculture (USDA). This is just recognition of the accreditation procedures; the respective certification bodies still have to meet the requirements of NOP to issue certificates accepted by the US (Huber and Schmid, 2009).

The Japanese Agricultural Standard (JAS) System established by the Japanese Ministry of Agriculture, Forestry & Fisheries (MAFF) was introduced in 1950 and it governs all the agricultural and forestry products, except for liquors, drugs, quasi-drugs and cosmetics. The JAS System consists of the combination of the 'JAS standard system' and the 'quality labelling standard system'. It assumed its current status in 1970 with the addition of the quality labelling standards system. The JAS standards for the production methods are called 'specific JAS standards' and they include the organic food standard. As of April 1, 2001, all plant based organic products and materials exported to Japan must be compliant with JAS. They must have an organic JAS seal affixed as it allows consumers to recognize organic foods easily. The JAS standards for organic plants and organic processed foods of plant origin are established on the basis of the guidelines for the production, processing, labelling and marketing of organically produced foods which were adopted by the Codex Alimentarius Commission at FAO. Operators wishing to export in Japan should certify products through a Japanese RCO (Registered Certification Organisation) or a foreign RFCO (Registered Foreign Certification Organisation) that is registered at the Ministry of Agriculture in Japan.

Organic standards and certification in Asia are set up in response to import requirements of the major importing organic markets. Many governments established organic regulations in hopes of establishing recognition from the EU and USA. Production and processing standards in the region in general reflect external requirements rather than local production conditions. Organic regulations are now in place in China, Japan, Korea, Israel, Philippines and Taiwan for domestic markets, imports and exports. The Indian regulation currently applies only to exports (Wai, 2009).

Thailand and Malaysia have published voluntary national organic standards, and they operate government certification programs as well. Indonesia has published voluntary national organic standards and operates a government accreditation scheme for private certification bodies. Sri Lanka, Vietnam, Laos and Cambodia are finalizing their respective national standards. Israel and

India have established recognition agreements with the EU. 157 certification bodies are listed for Asia in The Organic Standards (TOS) Certification Directory 2008 (Wai, 2009). 140 out of 157 certification bodies are found in just five countries: Japan (60); South Korea (32); China (29); India (13) and Thailand (6) representing the most important countries in Asia if we consider the market size or the organic land use size. Not all listed certification bodies are active. In spite of the fact that growth rate of the organic market in Asia is very high (between 15 and 20%: Sahota, 2009) most exports are certified by international certification bodies working in the regions. Local certification bodies in general are relatively weak and have small or no market share. Few local private certification bodies have established themselves outside of the regulated markets.

The interesting policy development among Asian decision makers emerged at the 'Organic Asia – the way forward' conference in Sarawak, Malaysia in 2008 is that the acceptance of imports should be on equivalence basis. The lack of harmonisation at international level represents a barrier to bilateral trade exchanges. The absence of harmonisation standards for conventional food is represented in this strand of literature as a factor representing transaction costs and reducing international food marketing (Henry de Frahan and Vancauteren, 2006). Some evidence particularly refers to organic food standards. Whereas scientific contributions show that harmonisation generates welfare improvements in the context of exchanges of products in specific organic markets (Wynen, 2004), Canavari and Cantore (2010) show that the existence of equivalence agreements for organic standards represents a signal of affinity in bilateral commercial relationships that is useful to identify those areas characterized by lower transaction costs in exchanges involving both organic and conventional food. It seems that a condition to grant equivalence agreements for organic products is that fruitful business relationships are already set up for conventional food. The consequent policy implication would be that discussions about harmonisation of organic standards can be driven only by considering marketing mechanisms concerning the whole agri-food market.

The importance of standards harmonisation and of equivalence of standards in international trading is therefore apparent. According to a definition of the European Commission (2001): '....*Harmonization* may be regarded as the drawing up of common or identical rules by a group of authorities, with the intention that the mandatory rules governing a product or service shall be the same among them' (EU Commission, 2001: emphasis in original). This definition implies that harmonisation is determined by effort produced by different countries in order to make similar rules and procedures of certification.

Equivalence is a mechanism to recognize and accept another system by acknowledging that variations between the systems uphold the respective systems objectives (WTO, 1994). With respect to conformity assessment, ISO defines equivalence as the sufficiency of different conformity assessment results to provide the same level of assurance. Equivalence therefore refers to achieving the same end, even though either the standard and/or the conformity assessment mechanism are not the same.

A regulatory system is composed of two crucial components: a 'rule', which can be a technical regulation, private standard or guideline against which a product or process is judged; and a 'conformity assessment system', which is a method and mechanism of assessing operators against the standard. The classical model of standards and conformity assessment establishes a framework for trade that is based on the principles of (1) harmonization and equivalence of standards and regulations and (2) mutual recognition of conformity assessment systems. (Courville and Crucefix, 2004).

Mutual recognition is a tool in which only the conformity assessment bodies are deemed to be equally capable and there is no attempt to converge the standards against which products are judged (Courville and Crucefix, 2004). It can be strengthened by activities implemented by international organisations such as IFOAM (International Federation of Organic Agriculture Movement) that are operating at global level to spread a common ground to promote organic farming rules and procedures. In 2004, 26 out of the approximately 360 certification bodies worldwide were IFOAM accredited and 22 of these had signed a multilateral agreement for mutual recognition (Bowen, 2004). The multilateral agreement for mutual recognition is useful when a party certified by one Accredited Certification Body (ACB) wishes to purchase a product certified by another ACB for use as an ingredient in a multi-ingredient product, or for re-sale. However we should underline that IFOAM accredited certification bodies currently represent less than 10% of the total world certification bodies as shown by Table 1.

Harmonisation, equivalence and mutual recognition are mechanisms that would reduce transaction costs in bilateral trade relationships. Our feeling is that all three aspects should be strongly encouraged by public authorities to reach the most promising results. It seems quite reasonable to think that the equivalence of organic standards could be more easily granted if decision makers acted for common targets and goals through harmonisation and if activities of certification bodies in each country were recognised abroad. Trustful trading relationships will be established when laws are perceived as 'similar' by operators at international level and when the activity of national certification bodies are accepted abroad because they conform to broadly accepted international rules.

However, we should point out that activities of each certification body should first be accepted and considered as credible, independent and objective in the domestic market as a preliminary condition to acting as leading organisations abroad. Certification bodies must tackle a trade-off in management activities: in a competitive market on one side they should care about the quality of inspections to preserve reputation, on the other side they tend to decrease audit quality to increase the market share with the aim to attract producers characterised by opportunistic behaviours. The typical scheme in a principal-agent model is represented by the relationship between the entrepreneur (principal) and the worker (agent). Theory shows that when the worker receives the wage and the entrepreneur cannot directly observe the performance of the workers, workers tend to perform less. In the relationship between the public authority (principal) granting the accreditation and the certification body (agent), the risk is that the performance of certification

Table 1. Certification bodies: approvals per region (Rundgren, 2009).

Region	Total	IFOAM	Japan	ISO 65	EU	USA
Africa	10	2	0	4	0	0
Asia	157	6	61	19	14	13
Europe	177	14	14	87	152	32
Latin America & Caribbean	48	5	4	17	6	10
North America	78	6	13	26	0	64
Oceania	11	4	6	4	7	5
Total	481	37	98	157	179	124

IFOAM: International Federation of Organic Agriculture Movements.
ISO: International Organization for Standardization.

bodies in terms of audit quality could be low if reputation is not affected or the costs for a lower reputation are negligible. A literature review on the evaluation and reliability of quality assurance systems is available in Albersmeier *et al.* (2009).

4. Organic certification in Italy: an attempt to analyse performance

Certification bodies in Italy are in charge of monitoring conformity of organic producers' processes and products to international law, with the aim of guaranteeing products with analogous features throughout the whole European Union territory.

They are committed to many duties other than verification and control of standard and procedures. They also play a political role because they connect public authorities and operators. Moreover, they have the responsibility to implement both the modalities by which international and national rules are applied and the management procedures for monitoring and control activities. Finally, they are charged to release to producers, on the basis of monitoring procedures, certificates allowing producers to use appropriate brands to signal products high quality. In Italy, the organic certification system is managed by 21 private bodies (16 certification bodies are Italian and operate in the whole Italian territory, while the other 5 certification bodies are not based in Italy and operate only in the autonomous province of Bolzano[4]). Public authorities monitor activities of third-party certification bodies.

We implemented a survey among the operators working in the 16 third-party certification bodies operating in the domestic market. Our questionnaire has been organised in two steps. The first step was aimed at identifying *structural factors* of certification bodies with a particular focus on those features affecting trust in the organic supply chain. The second part aimed at identifying *performance indicators* of third-party certification bodies. For the first step of the survey, the questionnaire was divided in four sections:
1. *General information* aimed at identifying the main structural and operational features of certification bodies. Through a qualitative approach we tried to obtain descriptive data about the characteristics of certification bodies' market activities.
2. *Supply chain relationships* aimed at identifying the nature of the relationship between third-party certification bodies and operators in the organic supply chain and its impact on the certification activities.
3. *Trust creation elements* aimed at identifying factors affecting trust in the third-party certification bodies.
4. *Reputation determinants:* in the basic economic literature reputation is defined as the set of attributes that agents associate to an organisation on the basis of its past activities (Weigelt and Camerer, 1988). In this study, however, we did not look at the factors of reputation but at the opinions of the other market agents. We wanted to know which agents were important and how important they were for the reputation of the third-party certifier. We interviewed third-party certifiers and applied the AHP (Analytic Hierarchy Process[5]) to calculate the importance weights for supply chain actors and their importance for reputation.

For the second step of the survey, another questionnaire was set up, focusing on collecting information about different indicators, aiming at describing certification bodies performance.

---

[4] The Bolzano province in Italy is inhabited by a German speaking minority, therefore it enjoys a high degree of autonomy and is regulated by specific laws.

[5] The AHP technique is a systematic method for comparing a list of alternative solutions and alternatives that is founded on mathematical grounds (Saaty, 1980).

In the next paragraphs we will briefly illustrate our results. In the first step of our analysis (structural factors of certification bodies) we interviewed 8 managers and 9 certification schemes managers working in 10 third-party certification bodies. The companies that accepted to participate into our survey represented 95% of the market at the time of the study. From the elaboration of the answers concerning the Section *General information* the most important features of third-party certification bodies' activity are: flexibility in the management of human resources, diversification of the supplied services and the SINCERT[6] accreditation. In particular accreditation of the third-party certification bodies can strongly influence the organic production system in terms of trust and reputation creation of the processes undertaken within the whole supply chain.

In the Section *Supply chain relationships* we identified main features of the relationships undertaken between certification bodies and external agents affecting the certification process. All experts agree on the fact that bureaucracy involving certification activities mainly creates inefficiency. Moreover, procedures and operational modalities for controls of public authorities towards certification bodies should be simplified and better clarified.

By the *Trust creation elements* results we identified a set of crucial factors for the creation of trust towards certification bodies. These factors were identified using direct interviews and have been divided into two groups: factors concerning *external visibility* (experience, no suspension of the authorisation by the competent public authority due to wrong certification activities and procedures, size of the certification body, other certification activities, SINCERT accreditation, accreditation for other organic certification schemes than the EU one, external activities) and factors concerning *internal operational modalities* (skills of the certification evaluating commission, management skills, skills of the certification schemes managers, clarity of procedures and documents, certification process management). According to previous literature about trust, all these factors can be identified as belonging to the cognitive dimension of trust rather than to the emotional one (Kramer, 1999; McKnight and Chervany, 1996). However, if we consider another important strand of research about trust creation factors (Castaldo, 2002; Kramer, 1999; Lewis and Weigert, 1985) we can interpret the trust cognitive factors emerging from our interviews as 'competences and skills'. Therefore, the initial classification into 'external visibility factors' and 'internal operational modalities' was abandoned. The identified factors are all considered as 'competences and skills', and further classified in 'basic skills', 'acquired skills' and 'further skills' as shown in Table 2.

Cognitive factors that create trust play a role when we introduce the concept of reputation. Concerning *reputation determinants*, we find that third-party certification operators identify different weights, according to whether they are managers or technical staff, to the importance of supply chain partners' opinions about the quality of their activity in the reputation formation process. Technicians of the certification schemes deem the opinions of public authorities the most important, whereas managers attribute a prominent role to the final customers' opinions (producers). Results are summarized in detail in Table 3. This finding is hardly surprising if we consider that technicians and managers play a different role in third-party certification bodies and reflect a different market perception according to their specific tasks: customers' satisfaction

---

[6] SINCERT (now ACCREDIA) is an Italian accreditation body (association of public and private stakeholders) aiming at monitoring third-party certification bodies activities and ability to implement with effectiveness certification procedures. All organic certification bodies must be authorized by the Italian Ministry of Agricultural, Food and Forestry Policies. Nine Italian certification bodies out of 16 are also accredited with SINCERT on a voluntary basis as a signal of the quality of their activity.

Table 2. Sharing of the trust creating skills identified by our interview in the 'basic skills', 'acquired skills' and 'further skills' groups.

Basic skills	Skills of the certification commission
	Management skills
	Technical employees skills
	Clarity in documents writing up
	Procedures management
	No suspension of the authorisation by the competent public authority due to wrong certification activities and procedures
Acquired skills	Years of experience
	Size of the certification organism
Further skills	SINCERT accreditation
	Supply of other certification schemes
	Non-EU organic certification standards
	External activities

Table 3. Evaluation of the importance of the supply chain operators' opinions for the Certification Bodies reputation.

	Weight (managers)	Weight (technical managers of the certification schemes)
Ministry	26.53%	39.14%
Regional authorities	18.45%	25.76%
Other certification bodies	9.82%	7.46%
SINCERT	15.20%	11.43%
Consumers association	9.40%	5.71%
Final customers	20.61%	10.50%

for managers and accomplishment of bureaucratic duties under public authorities' rules for technicians are, reasonably, the most important drivers.

In the second step of our analysis we analysed the *performance* of the third-party certification bodies according to different indicators. Experts identified number of audits, planning and timing of activities, sanctioning system for opportunistic behaviours and verifications features as crucial performance indicators. Four out of 16 certification bodies accepted to provide data about:
a. ratio between the number of inspections and the number of controlled farms;
b. ratio between the number of available products samples and the number of controlled products samples;
c. ratio between the number of punishment decisions and total controls;
d. ratio between the number of analysed samples and the number of audits;
e. ratio between the total number of sanctions to farms and the number of audits;
f. ratio between the number of announced audits and un-announced audits;
g. share of the cost for auditing personnel out of the total certification bodies costs;
h. ratio between hours of training and the number of inspectors.

The choice of these performance indices is based on the basic asymmetric information theory (Myerson, 1979). In a typical principal-agent model where the principal cannot directly observe the effort of the agent, the equilibrium effort level is lower than the optimal level. In this context, the solution to overcome this asymmetric information problem is to infer indirectly the agents' performance on the basis of observable variables such as the produced output.

We apply the same philosophy to evaluate the performance of the third-party certification bodies by collecting information about measurable performance variables representing the intensity and quality of the certification bodies effort to identify opportunistic organic producers' behaviour. Certification bodies could be induced to undertake collusive behaviours with producers to attract customers by superficial controls allowing farmers to sell products as organic without implementing the opportune practices and processes.

As a rough and non-conclusive proxy for certification bodies' performance, in this paper we assume that virtuous certification systems should show *the highest levels* of:
a.   number of inspections/number of controlled farms;
c.   sanctioning decisions out of the total controls;
d.   analysed products samples for each control;
g.   inspection costs out of the total certification bodies management costs;
h.   training hours/number of inspectors.

Virtuous certification bodies also should show *the lowest level* of the following indices, which have been constructed purposely to highlight less virtuous behaviours with higher scores:
b.   available samples/controlled samples (it is the reciprocal of the share, assuming a value 1 when all the available samples are controlled and growing rapidly when the share of controlled samples decrease);
f.   announced/un-announced audits (assuming a value 0 when all the audits are un-announced).

Results about these indicators for each certification body are standardised in a range [0,1], 1 being the maximum value of the indicator among certification bodies. This procedure allows comparability in a cross section and cross country dimension.

The aim of this study is not to define definitely the 'best' certification body on the basis of these indicators, but rather to investigate possible moral hazard actions. Bad performance indicators could signal the presence of opportunistic behaviour aimed at providing low quality and less intense inspections by reducing the number of controls, reducing inspectors' skills, or by allowing farmers to hide evidence of actions against the regular procedures for organic farming when they are aware of the date in which the inspection will arrive. Our results do not allow us to provide unambiguous evidence that a certification body is more virtuous than the others. In some cases, in fact, indicators seem to be quite contradictory. This contradiction could be caused by:
1.   The setting up of indicators is not fully insightful.
2.   Moral hazard behaviour of certification bodies can be expressed by different tools and is captured by different indicators.
3.   The relationship between certification bodies and farms/producers is not fully consistent with the principal-agent model. In other words, market failures persist because there is no objective indicator that can reveal producers' moral hazard. Moreover, diversity in the certification bodies' performance may be explained on the basis of different management and operations management practice rather than on the basis of opportunistic behaviours.

At this stage we are not able to go more inside through these issues but we deem that research directions devoted towards the empirical analysis of certification bodies behaviours is still less developed (see Anders *et al.,* in press) though theoretical contributions have existed for a few years (Jahn *et al.,* 2005). A reliability analysis of organic certification adopting a subjective measurement approach is performed by Albersmeier *et al.* (2009). They proceed analysing the attitudes of organic producers in Latin America and measuring the 'perceived usefulness', 'perceived cost', and an 'overall evaluation' of the organic certification. This may be seen as an alternative approach to the objective measurement approach we adopted in this study.

5. Conclusions

In this paper we focused on the activity of certification bodies in the organic supply chain in a broad perspective. In the first part, we mainly focused on a general description of the organic certification activities in different regions with a specific zoom on the harmonization and mutual recognition problems affecting international trade of organic food. We stressed how differences in regulations and the means by which they are enforced can generate market transaction costs. In the second part, we investigated for the specific case of Italy the performance of third-party certification bodies and we tried to provide some very preliminary findings about the objectiveness and independence of their work in a specific domestic market.

We divided our analysis in two parts. In the first part, we analysed the market role of certification bodies under different points of view: general features, opinions of the market agents' about certification bodies' activity, reputation determinants, factors affecting trust in the capability of third-party bodies to identify opportunistic behaviors of farmers and to guarantee products quality through transparent and independent procedures. We found that certification bodies play a crucial role in the organic supply chain to guarantee a correct information flow from producers to consumers through a quite complex environment. In this environment public monitoring authorities show superficial and inefficient control procedures and many internal skills and good relationships with external market agents are needed to consolidate the reliability of their activity. In the second part, we evaluated the performance of the third-party certification bodies by collecting information about measurable performance variables representing the intensity and quality of the certification bodies effort to identify opportunistic organic producers' behaviour.

However, many doubts arise regarding the actual performance of third-party certification bodies. Once identified a set of indicators expressing quality, intensity and frequency of inspections we found a great heterogeneity in the modalities by which certification bodies operate.

The drawback of our analysis is that we cannot truly identify the nature of these differences. In other words, we cannot easily understand if these differences can be interpreted as structural management discrepancies or as the outcome of opportunistic behaviors aimed at attracting customers through less tight control procedures. The finding is further complicated by the fact that we do not find results showing that a certification body can be considered more 'virtuous' than the others over time.

However, we deem that our work can represent an useful stimulus to enhance research activities to understand in depth certification bodies activities and further instruments to improve their role in guaranteeing the required quality level. In this paper we implemented an introductory approach to better understanding the context in which certification bodies operate, but further efforts are needed to expand and clarify connections between the organic supply chain tiers and factors that seem more promising to further enhance the organic sector performance.

Discussion in lecture room

Apply your theoretical knowledge with respect to this paper:
- Read the article from Wathne and Heide (2000) *Opportunism in Interfirm Relationships: Forms, Outcomes, and Solutions*. The article gives an overview on cost and revenue effects of opportunistic behaviour and summarises strategies to cope with opportunistic behaviour. Discuss which strategies favour which different actors in the supply chain such as governmental institutions or retail organisations.
- Moral hazard and opportunistic behaviour are 'positively' influenced by information asymmetries. The principal-agent theory distinguishes four different kinds of asymmetries; some are relevant before and some after signing the contract with an actor. Read about principal-agent theory and discuss the relevance for certification systems in the organic food supply chain.
- The asymmetric information theory claims that appropriate market signals are tools by which consumers can recognise high quality food products. Read about the adverse selection phenomenon and discuss the consequences of a poorly-functioning organic certification system. What are the best solutions to guarantee an appropriate performance of certification bodies? Who should control the controller?
- This paper uses the principal-agent theory to represent the public authority as principal and the certification body as the agent. Can the principal-agent theory be used to explain the relationship between the certification bodies and farmers? How could the performance of organic producers be observed?
- For further reading please refer to: M. Katz and H. Rosen, 1991. *Microeconomics*.

Acknowledgments

This paper is based on research partly performed in the framework of the BEAN-QUORUM project, co-funded by the European Commission through the Asia-Link Programme and in the framework of the Italian High National Interest Research Program (2004) 'Rural development, modern retail, food safety: organic farming perspectives in Italy'. The individual contribution may be specified as follows: Maurizio Canavari wrote the Introduction and the Section 1, Nicola Cantore wrote Section 2, Erika Pignatti wrote Section 3, Roberta Spadoni supervised the analysis and wrote the conclusions.

References

Albersmeier, F., Schulze H. and Spiller, A., 2009. Evaluation and reliability of the organic certification system: perceptions by farmers in Latin America. Sustainable Development 17: 311-324.
Anders, S., Souza Monteiro, D. and Rouviere, E., 2010. Competition and Credibility of Private Third-party Certification in International Food Supply. Journal of International Food & Agribusiness Marketing 22(3-4): in press.
Bowen, D., 2004. Current mechanisms that enable international trade in organic products. In: J. Michaud, E. Wynen and D. Bowen (eds.), Harmonisation and equivalence in organic agriculture. Volume 1. Background papers of the International Task Force on Harmonization and Equivalence in Organic Agriculture.
Canavari, M. and Cantore, N., 2010. Equivalence of organic standards as a signal of affinity: a gravity model of Italian agricultural trade. Journal of International Food & Agribusiness Marketing (in press).
Castaldo, S., 2002. Fiducia e relazioni di mercato. Il Mulino, Bologna.

Courville, S. and Crucefix, D., 2004. Existing and potential models and mechanisms for harmonisation, equivalency and mutual recognition. In: J. Michaud, E. Wynen and D. Bowen (eds.), Harmonisation and equivalence in organic agriculture. Volume 1. Background papers of the International Task Force on Harmonization and Equivalence in Organic Agriculture.

Cranfield, J., Deaton, J. and Shellikeri, S., 2009. Evaluating consumer preferences for organic food production standards. Canadian Journal of Agricultural Economics 57: 99-117.

Darby, M.R. and Karni, E., 1973. Free competition and the optimal amount of fraud. The Journal of Law and Economics 16: 67-88.

EU Commission, 2001. Commission staff working paper. Implementing policy for external trade in the fields of standards and conformity assessment: a tool box of instruments. Brussels 28.9.2001. SEC(2001) 1570.

Giannakas, K., 2002. Information asymmetries and consumption decisions in organic food product markets. Canadian Journal of Agricultural Economics 50: 35-50.

Henry de Frahan, B. and Vancauteren, M., 2006. Harmonisation of food regulations and trade in the Single Market: evidence from disaggregated data. European Review of Agricultural Economics 33: 337-360.

Huber, B. and Schmid, O., 2009. Standards and regulations. In: H. Willer and L. Kilcher (eds.), The world of organic agriculture: statistics and emerging trends. Fibl -IFOAM Report.

Jahn, G., Schramm, M. and Spiller, A., 2005. The Reliability of Certification: Quality Labels as a Consumer Policy Tool. Journal of Consumer Policy 28: 53-73.

Katz, M. and Rosen, H., 1991. Microeconomics. Richard D. Irwin, Inc. Illinois, USA.

Kramer, R., 1999. Trust and distrust in organizations: emerging perspectives, enduring questions. Annual Review of Psychology 50: 569-598.

Lewis, J.D. and Weigert, A., 1985. Trust as a social reality. Social Forces 63: 967-985.

McCluskey, J.J., 2000. A game theoretic approach to organic foods: an analysis of asymmetric information and policy. Agricultural and Resource Economics Review 29: 1-9.

McKnight, D.H. and Chervany, N.L., 1996. The meanings of trust. Working Paper, Carlson School of Management, University of Minnesota. Available at: http://misrc.umn.edu/wpaper/WorkingPapers/9604.pdf. Accessed June 2009.

Myerson, R., 1979. Incentive compatibility and the bargaining problem. Econometrica 47: 61-74.

Nelson, P., 1970., Information and consumer behavior. Journal of Political Economy 78: 311-329.

Rundgren, G., 2009. More than a million farms certified by 481 Certification Bodies. In: H. Willer and L. Kilcher (eds.), The world of organic agriculture: statistics and emerging trends. Fibl – IFOAM Report.

Saaty, T., 1980. The Analytic Hierarchy Process: Planning, Priority Setting, Resource Allocation. MC Graw Hill.

Sahota, A., 2009. The Global Market for Organic Food & Drink. In: H. Willer and L. Kilcher (eds.), The world of organic agriculture: statistics and emerging trends. Fibl –IFOAM Report.

UNCTAD/FAO/IFOAM, 2008. International Requirements for Organic Certification Bodies (IROCB), International Task Force on Harmonization and Equivalence in Organic Agriculture (ITF). Available at: http://www.itf-organic.org/irocbguide.html. Accessed August 2009.

Varian, H., 1987. Intermediate microeconomics: a modern approach. Norton. New York, USA.

Wai, O., 2009. Organic Asia - From Back to Nature Movement & Fringe Export to Domestic Market Trend. In: H. Willer and L. Kilcher (eds.), The world of organic agriculture: statistics and emerging trends. Fibl-IFOAM Report.

Wathne, K.H. and Heide, J.B., 2000. Opportunism in Interfirm Relationships: Forms, Outcomes, and Solutions. Journal of Marketing 64: 36-51.

Weigelt, K. and Camerer, C., 1988. Reputation and corporate strategy: a review of recent theory and applications. Strategic Management Journal 9: 443-454.

WTO, 1994. Agreement on Technical Barriers to Trade. Geneva, Switzerland.

Wynen, E., 2004. Impact of Organic Guarantee Systems on Production and Trade in Organic Products. In: J. Michaud, E. Wynen, and D. Bowen. Harmonisation and equivalence in organic agriculture. Volume 1. Background papers of the International Task Force on Harmonization and Equivalence in Organic Agriculture.

Part 2.
Understanding the value chain of quality food in east and west: selected case studies

Strengthening the export capacity of Thailand's organic sector

A. Kasterine[1], W.W. Ellis[2] and V. Panyakul[3]
[1]International Trade Centre, Rue de Montbrillant 54, 1202 Geneva, Switzerland
[2]GTZ, Thai-German Programme on Enterprise Competitiveness, 193/63 Lake Rajada Office Complex (16th floor), New Ratchadapisek Road, Klongtoey, Bangkok 10110, Thailand
[3]Green Net, 6 Soi Piboonupatam-Wattana Nivej 7, Suthusarn Road, Huay-Kwang, Bangkok 10310, Thailand

Abstract

Organic markets in developed countries are growing at an unprecedented pace, and in 2007 the global market was 46 US$ bn. Yet despite the acknowledged potential for Asian producers and exporters, the rate of farm conversion to organic systems cannot keep pace with demand, thus constraining the capture of economic value in global, domestic and intra-regional markets. Various technical assistance projects supported by the European Union, the UN International Trade Centre and others have identified a number of factors which act as underlying constraints to conversion, and called for regional cooperation to address these issues. This paper reports the findings and recommendations of a EU-funded technical assistance study on organic agriculture in Thailand, conducted by the UN International Trade Centre and Thailand's National Innovation Agency. The project was implemented from 2005 to 2006 primarily to offer national-level recommendations to strengthen Thailand's organic export sector and encourage compliance with EU regulations on organic imports, along with the requirements of private organic certification protocols. The project's recommendations addressed constraints identified during the stakeholder consultation and benchmarking processes, including research, production, training and extension, and the domestic and export markets. Drawing lessons from the project's national-level findings, the paper discusses emerging trends, challenges and opportunities for organic agriculture at regional level, noting a growing recognition of its significance and potential as a vehicle for sustainable livelihoods and poverty alleviation.

1. Introduction

Organic markets in developed countries are growing at between 20-30% a year, and in 2007 the global market was valued at US$46 bn. (Sahota, 2009). In the current global economic downturn, growth rates have slowed or reversed in some segments of organic products. With the apparent comparative advantages of Southeast Asian countries for organic production, there is potential for Asian producers and exporters to supply these key markets (Thode-Jacobsen, 2006).

In addition to traditional export markets like the EU, Japan and USA, important domestic and regional markets are also emerging in Asia. In China, India, Indonesia, and Thailand for example, the emergence of an affluent, health-conscious middle class, with changing tastes, rising health consciousness, and increasing disposable income is already driving a healthy demand within the subregion, and creating viable domestic markets for organic and other high-value specialty products (Wai, 2008). Southeast Asian countries are also ideally placed to serve the high-value markets in Japan and Korea.

According to a survey conducted by Global Industry Analysts, (RNCOS, 2009) the organic food and beverage market is predicted to surpass 70.2bn US$ (€49.4 bn) by 2010, with consumers increasingly turning 'organic' in search of health and safety in their food products. The major push towards healthy living has been felt most dramatically in Europe and the US, with both countries holding 80% of the overall world market. In the US, sales in the category are predicted to reach over 43bn US$ (€31bn) by 2010. However, it is the Asia-Pacific region that is the fastest growing region, posting a 28% compound annual growth rate.

On the supply side, there is ample underutilized agricultural land within the region, especially in upland areas, where pesticide use is minimal, and which may be ideal for establishing certified organic production zones without the need to pass through a long transition period before certification is granted. As an environmentally-friendly production system, organic systems are well suited to fragile upland agro-ecosystems, where heavy pesticide use poses occupational health hazards for untrained workers, as well as environmental risks.

Yet, despite triple-digit growth in the rate of farm conversion in Asia, supplies are failing to keep pace with global and regional demand (Organic Monitor, 2006). Constraints to conversion include lack of land tenure, inadequate access to technical training, information and support mechanisms, farmers' perception of risk, and high compliance costs. In economic terms, this gap means that opportunities for increasing organic exports are not being captured, and in environmental terms, there are continuing risks to natural resources arising from current agricultural practices.

With continuing consolidation of agri-food supply chains and increasing control by local and multinational corporates (Brown, 2005; Francis *et al.*, 2006; Vorley and Fox, 2004,), smallholders in Asia are facing formidable barriers to participation (FAO, 2004; Weinbergcr and Lumpkin, 2005). Stringent importing country requirements as well as private standards are transforming relationships within the supply chain. With modern trade retailers accounting for around 70% of global organic produce sold in 2005 (Asian Institute of Technology, 2005), closed supply chains based on contract farming and managed by large corporate operators are increasingly the preferred option to ensure year-round consistency of supply and compliance with these stringent standards.

One effect of such consolidation is to reinforce existing inequities in power relations within the supply chain, creating barriers to participation for organic smallholders in both export and domestic markets. This can be seen in the consolidation of production operations, and the smaller numbers of independent smallholders converting to certified organic methods.

On the regulatory front, Asian countries are at differing stages in developing standards and regulations for organic agriculture. International harmonization initiatives face some resistance, due at least in part to the competitive relationships between countries in the region (UNCTAD, 2006). The lack of mutual recognition of standards may thus act as a continuing constraint to growth in intra-regional trade.

Acknowledging these challenges as well as the potential of organic agriculture as a development tool (IFAD, 2005; Setboonsarng, 2006) the international donor community has become increasingly engaged in market development for organic agriculture. The European Union and the UN International Trade Centre (ITC) have initiated several country-based activities, notably their recent support via the Small Project Facility (SPF) (www.deltha.cec.eu.int/spf) and Asia Trust Fund (ATF) mechanisms (www.intracen.html)org/atf), respectively, for projects in Vietnam, Cambodia and Thailand.

From 2005-2006, the EU and the International Trade Centre (ITC) in partnership with Thailand's National Innovation Agency (NIA) implemented a technical assistance project 'Strengthening the export capacity of Thailand's organic agriculture' The project identified a range of challenges to development of both the domestic and export markets, and generated a series of national-level recommendations. The project and its recommendations are discussed later in this paper.

2. Objectives

The project had the following objectives:
1. To develop an innovative national organic model (strategy) for organic agriculture, whereby a consensus is reached regarding optimal allocation of public and private sector resources to support growth of the sector.
2. To facilitate the coordination of relevant government agencies in the implementation of organic projects.
3. To strengthen Thailand's government control system and requisites to prepare for application for inclusion in the EU's 'Third countries list' (Article 11 of EC Regulation 2092/91), the direct channel for EU member countries to accept imports of organic products from Thailand. Inclusion on this list would provide an important boost to Thailand's organic exporters.

3. Methodology

The project was implemented from July 2005 to June 2006 in three stages: (1) a benchmark survey supported by a literature review, stakeholder interviews and a National stakeholder workshop; (2) drafting of a National action plan for organic agriculture in Thailand, with a particular focus on facilitating exports of high quality organic produce from Thailand to the EU; and (3) a series of specialist training workshops on information and skills required to strengthen government control systems for organic agriculture. In addition, private sector representatives received training in requirements for compliance with EU legislation. The study methodology and stakeholder consultation process are shown schematically in Figure 1.

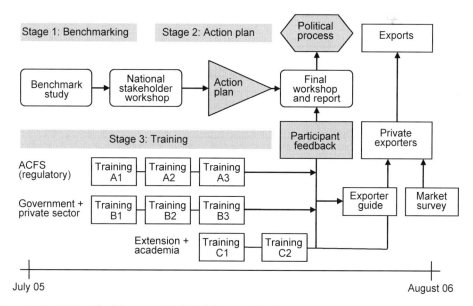

Figure 1. Project methodology and stakeholder consultation.

The outputs of the consultation process were consolidated into a National action plan (ITC 2006), which was presented to the stakeholder community and senior government officials at a National round table, held on 30 August 2006 in Bangkok.

4. Organic agriculture in Thailand

Thailand has long been a major exporter of tropical fruit and vegetables to European markets and is recognized as a reliable source of high quality produce. The country's organic sector is small but has grown rapidly over the past five years in line with global trends; the certified organic land area increased by more than 90% between 2001 and 2005, to 21,701 ha (Green Net, 2006), (Figure 2). This growth is attributed mainly to growing consumer consciousness, declining incomes from conventional farming, and environmental concerns. Nevertheless, Thailand's certified organic area, at below 1% of total agricultural land, is among the lowest in the world (Willer *et al.*, 2006) and remains insufficient to meet increasing demand for export and, increasingly, at home.

In 2007 the total volume of organic products in Thailand delivered to market was estimated at 33,678 tons, valued at 19.5 million euros: a substantial increase from just 9,756 tons/7.5 million euros in 2003 (Table 1).

The largest production category is organic rice, primarily *Hom-mali* jasmine rice from the northeast region, followed by fresh vegetables and herbs. Organic vegetables are mainly leafy vegetables, especially the salad type and Chinese vegetables, produced mainly in central Thailand and in Chiang Mai province. Dedicated organic orchards are also becoming more important, though many organic vegetable farms also produce organic fruits. The major fruits now grown organically are mango, papaya, and longan.

According to a study done by the International Trade Centre (ITC), commissioned for the project (Thode-Jacobsen, 2006), Thailand has considerable potential in producing and exporting

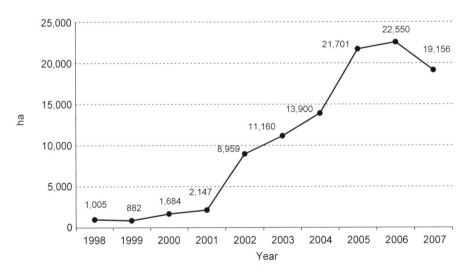

Figure 2. Organic agriculture growth in Thailand 1998-2007 (Green Net/Earth Net Foundation, 2008).

Table 1. Production and market value 2003-2007[1] (Green Net/Earth Net Foundation, 2008).

Crop	2003			2005			2007		
	Production (tons)	Value		Production (tons)	Value		Production (tons)	Value	
		m baht	m euros		m baht	m euros		m baht	m euros
Rice	7,007.9	210.24	4.20	18,960	534.8	10.7	13,467	373.40	7.5
Field crops				2,041	45.2	0.9	2,934	65.90	1.3
Vegetables and herbs	2,671.3	160.28	3.21	4,618	255.8	5.1	5,337	297.20	5.9
Fruits				3,747	74.9	1.5	11,930	236.60	4.7
Others	76.9	4.61	0.09	49	9.7	0.2	9	1.80	0.0
Total	9,756.1	375.1	7.5	29,415	920.4	18.4	33,678	974.90	19.5

[1] Exchange rate: baht/euro: 50.0, 2005.

organic products (both basic and processed) to EU member countries. The study found that dried fruits, canned fruits and vegetables, fruit and vegetable juices, and processed cereal foods have greatest market potential in this respect. Non-food organic products too (e.g. clothing, toiletries, decorative and spa products) are also of considerable interest.

However, despite dynamic supply side growth in recent years, food exporters face major challenges in realizing this potential due to the need to comply with the EU's stringent food safety and organic certification regulations e.g. the European Commission Council Regulation (EC) No. 834/2007 and especially No. 1235/2008, regulating the imports from third countries.

5. The supply chain

Thailand's organic products are largely produced by private sector operators and by projects supported by NGOs. In the private sector, farmers either work on large farms, or operate as contract farmers at pre-agreed volumes, grades and price levels. In the contract farming arrangement the companies typically provide technical advice and cash advances to the contract growers. Generally, the company will also pay for and hold the organic certification in their own name. The costs of conversion and ongoing inspections are also mostly covered by the company which can access the necessary knowledge, capital and technical resources (ITC, 2006).

In projects supported by NGOs, producers normally work together as a grower group, although few have formal registration as cooperatives or farmer organizations. They are self-supporting and provide their members with technical training, cash advances to cover production costs, and collective marketing mechanisms. The Green Net organization exemplifies such an arrangement.

Currently the main distribution channels in Thailand are the member system (weekly deliveries to subscribed customers within the community), weekly markets, retail health shops, health supermarkets, supermarkets and organic restaurants. The majority of produce is sold through supermarkets, which by now have superseded health shops or 'green shops' in importance.

Surveys indicate that consumers are ready to pay a maximum premium of 10-15% for organic produce. As shown in Table 2, organic produce attracts a significant but highly variable retail premium over conventionally produced vegetables. However, as more organic agricultural produce is launched in the market, premium levels are expected to decline and may even disappear.

Despite declining farm-gate premiums, and the increasingly weak position of smallholders to negotiate fair prices for produce with large supermarket chains, still there remains sufficient price differential for growers as an incentive for conversion. The price differential at farm level is clearly demonstrated in the case of organic asparagus. Growers obtain an average price of 32 baht/kg (euros 0.64/kg) for organic asparagus in comparison to 22-25 baht/kg (euros 0.44-0.50/kg) for conventional asparagus (Boonyanopakun, 2007: personal communication). A member of the Srakaew contract farming arrangement with two rai (0.32 ha) of farm land is able to generate an average income of 132,130 baht (euros 2,643) per year from organic asparagus (Uathavikul, 2004) in comparison to an average income of 115,000 baht (euros 2,300) per year obtained from conventional asparagus (Department of Agriculture, 2007). The differences are summarised in Table 3.

In 2003, members of a Srakaew organic asparagus contract farming operation were able to generate an average annual income of 132,130 baht (euro 2,643) from two rai of farm land (equivalent to euros 8,258/ha/yr) (Uathavikul, 2004). A survey at the same site also showed that the cost of farm management using organic methods was 40-59% lower than the cost of conventional methods. A steady stream of income, combined with lower variable production costs, can make organic farming an attractive choice for growers.

Table 2. Price comparison of fresh vegetables in Bangkok supermarket (baht/kg) (Earth Net Foundation/Green Net, 2008).

	Conventional vegetables	'Hygienic vegetables'[1]	Organic vegetables	Organic premium (compared with 'hygienic vegetables')[1]	Organic premium (compared with conventional vegetables)
Dec 01	40.18	54.79	88.375	61%	120%
Jul 02	41.28	60.28	66.99	11%	62%
Mar 03	38.98	35.29	65.94	87%	69%
Dec 03	57.49	56.75	64.28	13%	12%
Mar 04	29.45	46.29	76.77	66%	161%
Aug 04	34.85	83.3	135.29	62%	288%
Feb 05	29.85	48.24	52.37	9%	75%
Aug 05	49.29	64.22	45.84	-29%	-7%
Apr 06	46.03	75.38	108.03	43%	135%
Jul 07	43.88	103.06	134.17	30%	206%
Dec 07	42.13	105.26	122.80	17%	191%

[1] Hygienic vegetables are produced using good agricultural practice.

Table 3. Price comparison at farm level (per 2-rai family plot) (Department of Agriculture Thailand, 2007: available at www.doa.go.th/en/; Swift Co. Ltd, 2006: personal communication).

Variable	Organic asparagus	Conventional asparagus
Price per kg (baht[1])	32	22-25
Yield (kg)	2,064	2,405
Annual income per two rai[2] of land (baht)	132,130	115,000
Annual income per hectare (euro)	8,258	7,188

[1] Exchange rate: baht/euro: 50.0, 2005.
[2] Rai= 0.16 hectare.

6. Challenges and constraints identified during the stakeholder consultation process

The analysis revealed many problems in organic projects initiated by government, the private sector and the NGO community in Thailand. Key issues are listed as follows:
- Lack of assistance for farmers during the conversion period, which can take 1-3 years. Small farmers in particular have insufficient capital, knowledge and resources to risk converting to organic farming if they must carry the high compliance costs.
- Following conversion, little technical support is provided.
- There have been relatively few advances in soil improvement technologies and crop protection technologies while the basic concept of organic farming – as a positive farm management system, with its broader philosophy of attempting to conserve and rehabilitate the agro-ecosystem – is often overlooked.

There is insufficient education and competency development to enhance capacity at producer, processor and exporter levels to better manage the production process. The consultation process also revealed that existing organic farming systems often cannot adequately address the fundamental problem of ensuring consistent production and regular supply of fresh produce of guaranteed quality. Production and product quality are vulnerable to fluctuations in biotic factors and unpredictable changes in growing conditions and the processing methods remain unsophisticated. Often, processing is managed and operated at community level, and relatively little progress has been made in improving post-harvest technologies to minimize post-harvest losses.

As a result of the failure to improve the capacity of producers, many organic initiatives would not qualify as meeting internationally-recognized organic standards. Moreover, the organic guarantee system is not fully understood even by organizations promoting organic agriculture. In particular accreditation and certification are frequently not properly differentiated, and regulations covering organic imports are not well understood even amongst practitioners. The findings also showed that overall, the level of government and private sector investment in advertising and promotion is inadequate. There are few media channels directly providing information on organic agriculture, and few entrepreneurs invest in paid advertisements.

7. Recommendations and national action plan

In response to the above findings identified during the stakeholder consultation process, a series of recommendations were developed as a national action plan. The plan aimed to dovetail with ongoing initiatives under Thailand's national organic agenda, and was the result of extensive talks with growers, regulators, companies and researchers. In all the plan contains seven policy recommendations to strengthen the sector, covering production, regulations, certification, research, training and marketing. In support of each recommendation, specific actions were proposed to support both the export and domestic markets. The project's final report including full recommendations are summarized in Table 4, and are available in full at Intracen (www. intracen.org/Organics/documents/Strengthening_the_Export_Capacity_of_Thailands_ Organic_Agriculture.pdf).

It is hoped that if successfully implemented, these measures will help the sector become more efficiently organized, ensuring a wider range of produce for both domestic and export markets, with supplies matched more closely to demand. However, this can only be realized through closer linkages between producers, exporters and the overseas markets, to ensure understanding and compliance with importer protocols and national standards.

Underlying the national recommendations are two principles which are more broadly applicable. First, the success of organic agriculture depends upon the capacity and competency of private sector actors who must play a key role in its development. Governments should thus be encouraged to play an enabling and facilitating role, establishing effective and transparent mechanisms, which are internationally recognized. The government's responsibility should be to support and oversee the private sector, help to open up new markets (domestic and exports) and uphold national standards as well as international obligations. In this respect, public-private partnerships can often result in more workable and sustainable solutions.

Secondly, the recommended interventions will be most effective in generating long term sustainability if they can be implemented at the community level (i.e. bottom-up). Prioritization of training, research, accreditation and support for farmers during the conversion period will also serve to stimulate conversion (particularly for smallholders), broadening and diversifying the production base. Again, close consultation with, and participation of the private sector and non-governmental organizations will help achieve effective long term solutions.

8. Discussion

The Thailand analysis indicated a strong potential for export of organic products, both food and non-food. Apart from presenting attractive market prospects for fresh organic produce, there are additional avenues to stimulate local economies through:
- local value-added through vertical integration with processing and on-site retail packing;
- development of organic production zones linked to eco-tourism to present off-farm income-generating opportunities;
- product innovation (e.g. herbal products, and non-food products such as clothing, toiletries, and spa supplies).

Despite these opportunities, technical, economic, structural and political constraints continue to hinder market development. The project concluded that there is a need to establish appropriate mechanisms to improve information flows (a) among stakeholder groups; (b) between importing and exporting countries; and (c) among countries within the region itself. Alliances (whether

Table 4. Recommendations for a national action plan (Thailand).

Strategy 1:	Broaden the production base for organic agriculture
Action 1.1	Implement additional support measures to facilitate conversion to organic systems
Action 1.2	Support the establishment of organic production clusters in the private sector
Action 1.3	Support contract farming in organic agriculture as an effective vehicle for poverty alleviation
Action 1.4	Invest in technologies and processing facilities to enhance value-added and exploit new market opportunities
Action 1.5	Support the organization of growers in regard to joint distribution, storage and transport infrastructure
Action 1.6	Strengthen the ongoing bio-fertilizer initiative spearheaded by the Ministry of Agriculture and Cooperatives
Strategy 2:	Enhance capacity and streamline the existing regulatory structure
Action 2.1	Review the public sector certification system and improve access by smallholders
Action 2.2	Review and strengthen the voluntary national organic standards to improve understanding and enhance their value to farmers
Strategy 3:	Prioritize research into organic agriculture
Action 3.1	Identify and address the role and potential contribution of organic agriculture to national goals for sustainable development
Action 3.2	Establish a national organic research and development centre and national organic information database
Action 3.3	Earmark additional funding for multidisciplinary research in order to address key challenges
Action 3.4	Encourage researchers to examine and evaluate traditional knowledge about pest control treatments, working in close collaboration with farmers and local communities
Strategy 4:	Enhance and upgrade training and extension services for organic farmers
Action 4.1	Promote organic agriculture through a participatory community-level approach
Action 4.2	Initiate and support training programs for farmer groups to help them set up internal control systems as further options to reduce compliance costs for smallholders
Strategy 5:	Develop the domestic market for organic goods
Action 5.1	Conduct market research in order to understand consumer preferences and behaviour
Action 5.2	Private sector stakeholders should strengthen their representation through greater participation and support for the Thai Organic Traders' Association
Action 5.3	Introduce a pro-organic public procurement policy by public agencies
Action 5.4	Establish an effective market information system for organic produce
Action 5.5	Initiate public awareness campaigns to stimulate demand and promote consumption
Strategy 6:	Expand the export market for organic goods
Action 6.1	Extend additional support for exporters through global marketing outreach initiatives, liaison and export facilitation processes
Action 6.2	Review and maximize potential of innovative marketing channels for organic produce
Action 6.3	Provide an effective global market information service for organic exporters
Strategy 7:	Establish Thailand as a leader and centre of excellence in organic agriculture at the regional level
Action 7.1	Lead initiatives to foster cooperation between governments in Asia on harmonization of national regulatory regimes and sharing of experiences on key issues
Action 7.2	Foster regional collaboration among private-sector certification bodies
Action 7.3	Develop training courses for organic conversion schemes at regional level
Action 7.4	Establish a regional organic trade association

formal or informal) between organizations across the region could thus lead to synergies and enhanced competitiveness of the overall sector. This is especially true for certification, which has not yet evolved to establish stable nationally-based services, or – as mentioned above – mutual recognition of inspection services.

Such integration and mutual recognition may be a logical next step towards a harmonized certification regime. Indeed, the International Task Force on Harmonization and Equivalence in Organic Agriculture, convened by FAO, IFOAM and UNCTAD, has been working since 2003 as an open-ended platform for dialogue between public and private institutions (intergovernmental, governmental and civil society) involved in trade and regulatory activities in the organic agriculture sector (Rundgren, 2006). The ITF's objective is to facilitate international trade and access for developing countries to international markets.

In terms of the government, its role should be seen as a facilitating one, with its priority first and foremost to support private initiatives to broaden and diversify the certified production base and minimize certification and other compliance barriers (technical and cost-wise) for smallholders. Furthermore, action is needed to ensure clear separation between government's roles in accreditation and certification, and to encourage the development of a Thai-based private sector certification industry.

9. Regional implications and conclusions

Since many Southeast Asian countries face broadly similar constraints (research, production, standards and market development) to those identified for Thailand, many of the issues raised by the Thailand case study also resonate at the regional level. Examples include the need for greater coordination among stakeholders, and for government support measures to reduce risks during conversion periods and strengthen the supply base to ensure stable, consistent supplies of a wide range of quality organic produce to serve regional and international markets. There are also parallel implications for standards and national control systems. UNCTAD/UNESCAP (2006) recently highlighted the need for regional collaboration to promote intra- as well as inter-regional trade in organic agriculture.

Unchecked, the increasing complexity of both public and private production standards, combined with uncertain policy environments in producing countries, are likely to exacerbate the plight of small farmers and contribute towards exploitative trading relationships and unsustainable use of natural resources. A *laissez-faire* approach to organic market development in Asia might thus be expected to lead to increasing marginalization of small farmers (FAO, 2004), thus posing a challenge to the organic movement's founding principles of sustainability and empowerment of small farmers.

Given that the organic sectors of Asian countries are (with some exceptions) at relatively early stages in their development, there is a real window of opportunity to support the development of sustainable markets and regulatory structures in a way that espouses the organic movement's founding principles, and encourages an appropriate balance between large-scale and smallholder operators. Such rationalization will be important not only in facilitating trade (e.g. on TBT/SPS issues) but also in fostering the sharing of 'best practice' technologies at producer level.

Such regional-level interventions can help governments lay down effective public policy frameworks that will allow organic supply chains and markets to capture the full value of natural resources in a sustainable manner, and promote the interests of poor people around Asia.

Discussion in lecture room

Apply your theoretical knowledge with respect to this paper:
- Challenges and constraints of organic farming in Thailand are due to a lack of education and training of farmers. Discuss different models of knowledge transfer such as the technology and innovation transfer model of Rogers (2003) *Diffusion of innovations*, or participatory approaches such as farm field schools, which are regularly applied in development projects.
- Read Sumberg (2005) *Systems of innovation theory and the changing architecture of agricultural research in Africa*. Discuss if the theoretical findings in this article are relevant to Thailand.
- In the year 2000 FAO published a guidelines and visions document for Agricultural Knowledge and Information Systems for Rural Development (AKIS/RD). Imagine you are responsible for building up a system for knowledge transfer to organic farmers encompassing education, research and extension. Search and download the FAO documents from the Internet and develop a concept for a sustainable extension system for organic farming in Thailand.
- As highlighted in this paper most farmers in the private sector of organic farming in Thailand either work on large farms, or operate as contract farmers. Discuss the ethical implications and if this situation is compatible with the 'organic philosophy'.

Acknowledgements

The financial and technical contributions of the International Trade Centre (UNCTAD/WTO), the European Commission and the National Innovation Agency, Thailand, are gratefully acknowledged.

References

Asian Institute of Technology, 2005. Research Report: An Analysis of the Trend and Situation of the Thai Organic Trade (in Thai), 2-71, Table 2:22.

Brown, O., 2005. Supermarket Buying Power, Global Commodity Chains and Smallholder Farmers in the Developing World. UNDP Human Development Report Office Occasional Paper.

FAO, 2004. The State of Food Insecurity in the World 2004. Food & Agriculture Organization of the United Nations, 22-24.

Francis, C., Koehler-Cole, K., Hansen, T. and Skelton, P., 2006. Science-Based Organic Farming 2006: Toward Local and Secure Food Systems. University of Nebraska – Lincoln, Extension Division, Center for Applied Rural Innovation. Available at: http://cari.unl.edu/Organic-Farming-2006.doc.

Earth Net Foundation/Green Net, 2008. Organic Statistics Thailand. Available at: www.greennet.or.th. Accessed January 2010.

ITC, 2006. Strengthening the Export Capacity of Thailand's Organic Agriculture Final Report. Prepared by E.W. Kasterine, A. Panyukul, V. Vildozo, D. ITC, Geneva. Available at: www.intracen.org/Organics/documents/Strengthening_the_Export_Capacity_of_Thailands_Organic_Agriculture.pdf. Accessed December 2009.

IFAD, 2005. Organic Agriculture and Poverty Reduction in Asia: China and India Focus Thematic Evaluation. International Fund for Agricultural Development. Report No. 1664, July 2005, 24-26.

Organic Monitor, 2006. The Global Market for Organic Food & Drink: Business Opportunities & Future Outlook. November 2006.

RNCOS, 2009. Emerging Organic Food Markets: Market research report, February 2009.

Rogers, E.M., 2003. Diffusion of Innovations. 5th edition, Free Press, Washington D.C., USA.

Rundgren, G., 2006. Best practices for organic marketing regulation, standards and certification: Guidance for developing countries. Draft document, UNCTAD International Task Force on Harmonisation and Equivalence in Organic Agriculture. Available at: www.unctad.org/trade_env/ITF-organic/meetings/itf6/ITF_Regulation_Guidance_Draft_GR060908.pdf. Accessed January 2009.

Sahota, A., 2009. The Global Market for Organic Food and Drink, In: H. Willer and L. Kilcher (eds.), The World of Organic Agriculture. Statistics and Emerging Trends. IFOAM, Bonn; Fibl, Frick; ITC Geneva, Switzerland, pp. 59-64.

Setboonsarng, S, 2006. Organic Agriculture, Poverty Reduction and the Millennium Development Goals. ADB Institute Discussion Paper No. 54.

Sumberg, J., 2005. Systems of innovation theory and the changing architecture of agricultural research in Africa. Food Policy 30: 21- 41.

Thode-Jacobsen, B., 2006. Market Intelligence Study: Market Opportunities for Organic Products from Thailand. Report to UN International Trade Centre by Bioservice Co Ltd, 2006.

Uathavikul, P., 2004. Poverty Reduction through Contract Farming. Lessons from Srakaew Province, Thailand: ADB-UNESCAP regional workshop on contract farming and poverty reduction, 9-12 August, Bangkok, Thailand.

UNCTAD, 2006. Study of the Potential Impact of Equivalence on Competition Among Operations. International Task Force on Harmonisation and Equivalence in Organic Agriculture. Draft Concept Note 2006.

UNESCAP, 2006. United Nations Economic and Social Commission for Asia and the Pacific: regional workshop on the trade and environment dimensions in the food and food processing industries in Asia and the Pacific, 16-18 October, Bangkok, Thailand.

Vorley, B. and Fox, T, 2004. Global Food Chains- Constraints and Opportunities for Smallholders. Prepared for the OECD DAC POVNET Agriculture and Pro-Poor Growth Task Team, Helsinki Workshop, 17-18 June.

Wai, Kung O., 2009. Organic Asia - From Back to Nature Movement & Fringe Export to Domestic Market Trend. In: H. Willer and L. Kilcher (eds.), The World of Organic Agriculture. Statistics and Emerging Trends, IFOAM, Bonn; Fibl, Frick; ITC Geneva, Switzerland, pp. 134-138.

Weinberger, K. and Lumpkin, T.A., 2005. Horticulture for poverty alleviation – the unfunded revolution. Shanhua, Taiwan: AVRDC – The World Vegetable Center, AVRDC Publication No. 05-613, Working Paper No. 15.

Willer, H., Yussefi-Menzler, M, and Sorenesen, N (eds.), 2008. The World of Organic Agriculture: Statistics and Emerging Trends 2008. International Federation of Organic Agriculture Movements, Bonn, and the Research Institute of Organic Agriculture (FiBL), Frick, Switzerland.

Factors influencing purchasing decisions of Austrian distribution channel operators towards 'made in China' organic foods

R. Haas[1], C. Ameseder[1] and R. Liu[2]
[1]*University of Natural Resources and Applied Life Sciences Vienna (BOKU), Institute of Marketing and Innovation, Feistmantelstrasse 4, 1180, Vienna, Austria*
[2]*Xinjiang Agricultural University, No.42 Nanchang Road, Urumqi, Xinjiang, China*

Abstract

Values & attitudes as well as the knowledge & awareness about an organic product influence consumer preferences and their purchase decision. Especially Austrian origin, regional products and environmental issues are among other factors important for Austrian consumers. However, production lags behind demand in Europe, the largest market for organic products worldwide; bottlenecks are one of the main problems of the sector since 2005. Consequently imports especially form third countries are getting more important. The following paper focuses on internationally produced organic products and their market opportunities in Austria, one of the most developed organic consumer market worldwide. Interviews with key players of the Austrian sector of organic food were conducted to assess their purchase criteria. Results from a qualitative survey demonstrate the decisive role of the retail and the industry as gatekeepers for the whole sector, considering values and attitudes of their customers.

1. Introduction and problem description

Global demand for organic food is still rising, worldwide sales are increasing by over five billion US$ a year. 2006 international sales have reached 38.6 million US$; double than that of 2000, when sales were at 18 billions US$. Still the consumer demand for organic products is concentrated in North America and Europe. 97% of revenues are realized in these two regions (see Willer *et al.*, 2008). Due to this continuous market growth the global organic food industry has been experienced acute supply shortage in almost every sector since 2005. Sahota (2008) expects therefore growth of the organic sector to be stifled. Already in the last years Asia, Latin America and Australasia took the chance and became important producers and exporters of organic foods. In recent years these countries already raised the proportion of organically managed land driven by the worldwide trend of harmonization of the agricultural market and production systems and the supply bottlenecks in Europe and North America. Especially China is promised to raise exports within the next years. Although the total value of Chinese organic export in 2004 was only estimated to be 350 million US$, organic farming in China has grown rapidly in terms of farm numbers, arable land and export value. Still most of the organic products sold domestically are reportedly sold without an organic premium, so the primary driver of Chinese organic food and farming is still trade and export. Primary export markets are North America, Europe and Japan. Parallel to the rapid growth in organic production and export, there has been a dramatic increase in the number of organic certification bodies in China (Kledal *et al.*, 2007). In 2004, six organizations offered certification services, and the year after this number had increased to 26 (USDA, 2006). But China not only set up institutions. Driven by the increasing sales in North America and especially in Europe, China gained economic and scientific capacity as well allowing the country to *become a significant player on the world market for organic foods* (see Kledal *et al.*, 2007).

The European market of organic food & drink products is the largest and most sophisticated market worldwide valued at about more than 20 billion US$ in 2006 (14.3 billion euros) and covering more than 50% of the worldwide sales volume (see Sahota, 2008). Within Europe Austria is an organic pioneer and hosts one of the most developed consumer markets worldwide of organic produce. Annual sales of organic products in Austria are estimated around 860 million euro (1.3 billion US$, Bio Austria 2007). The per capita consumption of 56 euro (74 US$) is one of the highest worldwide as is the market share of organic food, which is of about 6% of total food & drink sales (e.g. USA: per capita consumption of 34 euro (45 US$) and a market share of 2.5%; Germany: per capita consumption of 47 euro (62 US$) and a market share of 3% (see Haumann, 2007; Richter and Padel, 2007). 87% of Austrian consumers purchase organic food products at least occasionally (AMA Marketing, 2007). Reasons for the sustainable growth are besides the main purchase motives that organic food is perceived to be healthier, the diminishing price differences to conventional food and the strong distribution coverage of organic food through conventional retail channels. Considering all distribution channels (supermarkets, organic food stores, direct sales from farms, restaurants, communal feeding), supermarkets have by far the biggest market share of 64%. Especially REWE, the retail market leader, reports revenues of organic food sales by his own account of €220 million (around 320 million US$) in 2006, representing alone a market share for organic food products of 40%. In Germany for instance the situation is different because organic food stores have traditionally bigger market shares there. The high market shares of conventional retail in Austria are caused among other factors by the early market entry more than ten years ago (1994) by REWE into the organic food market and by a consequent brand building strategy. The private label 'Ja natürlich!' (Yes, naturally) from REWE is the best-known and strongest organic brand in Austria (Vogl et al., 2005). Organic food stores only have a market share of 14%, nevertheless these stores, including their wholesalers, are very important for the whole sector (Bio Austria, 2007: unpublished presentation). Compared to supermarkets product assortment is much bigger in these stores, (up to 10,000 products) entry barriers for new products are smaller in this distribution channel and specialized traders are important for opinion making in the branch.

Normally one would expect that product and brand innovation would come from the food industry sector. Not in the case of organic food. The reasons behind this development are for example that the organic food sector in Austria is highly fragmented and mostly run by small and medium sized companies, which lack the financial capacity and often the marketing know-how to develop strong organic brands (strong in the sense of brand awareness and brand identity). Secondly in the last decades the conventional food retail sector faced increasing competition, decreasing profit margins and saturated markets. In face of these challenges retailers had to develop new strategies to gain new customers. Experiences with the Ja natürlich brand (nowadays with 630 food products in 13 product categories) show that the retail chain acquired new customers and significantly improved its image by offering organic food.

As Europe, the Austrian sector of organic food faces the problem of acute supply shortages since 2005 (see Willer and Yussefi-Menzler, 2008). While the demand of organic food items is still rising, the production area measured by organic acreage in Austria stays almost constant, as Figure 1 shows. Acreage for the production of organic food cannot be increased that fast, as a changeover from conventional agricultural production to organic production in Austria normally takes two up to three years (Bio Austria, 2008: available at http://www.bio-austria. at/bio_bauern/umstellung). Furthermore organic yields are in average 30% less compared to conventional farming methods. Therefore the market in Austria changed in the last years *from a supply-oriented market to a demand-oriented market.*

The situation in Germany, the biggest market for organic food products in Europe is even more challenging. Demand for organic produce has increased faster than supply; several key products

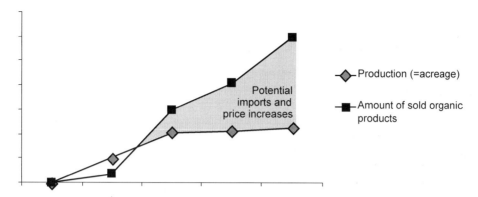

Figure 1. Organic farm acreage in Austria compared to volume sold (own calculation based on Invekos data and Bio Austria sales information).

are now undersupplied (Padel *et al.*, 2008). For instance demand for organic milk in retail shops increased in Germany by 35% in 2007 (Rippin *et al.*, 2007), which leads to supply bottlenecks not only in Germany but also in Austria. Austria is highly affected by these developments of the German market, as retailers and wholesalers especially from Germany purchase in Austria as well.

This increase of national demand of organic food products as well as higher demand of neighboring countries – especially from Germany – leads to a high potential for imports and price increases in Austria, as production in Austria more or less is stable. As mentioned above Asia, Latin America and Australasia are gaining importance in the worldwide market of organic foods, less for their home markets than for the prospective international export markets. But despite the increasing supply shortages critics could say that long transport routes for organic products are not coherent with the core organic philosophy of self-sufficiency and localness. Climate change and food production are issues getting more and more linked together in public awareness. For e.g. international retail chains invent labels such as 'air freight free' labels (Tesco) or NGO's calculate carbon footprints to compare conventional with organic food. Daily newspapers pose the question if organic strawberries on a supermarket shelf in Vienna during winter, imported from Spain over 2,500 km, make sense from an organic point of view (Brickner, 2007).

However, the future of organic agriculture will depend, to a large extent on consumer preferences and the resulting demand. Therefore, when talking about international trade of organic food factors influencing consumer preference and purchase decision have to be considered first of all.

2. Objectives and research questions

As mentioned before import potential for organic products in Austria and Germany is given, as production lags behind demand in whole Europe and North America. China, as well as some other countries, is about to become a significant player on the world market of organic foods. Nevertheless imports from China face two problems, first long transport routes, which could be negatively seen by critical consumers and secondly consumers could distrust the fulfillment of organic production and processing standards in China. How trustworthy are Chinese or international certification companies controlling the organic production process in China? Putting these factors together raises crucial marketing questions:
- Is there a positive or negative country of origin effect towards organic products from China?

- Do Austrian distribution channel operators accept organic food products with long transport routes?
- Which organic food products are mainly purchased from China?
- Which product categories show higher market potential for Chinese organic products?
- What are the purchase criteria concerning international organic products of key players in the branch?

So in answering these questions the aim of this study is to give an overview of the market potential for organic food products from China.

3. Factors influencing purchase decisions of consumers

Most consumers purchase organic products because of a perception that these products have unique (and in some cases superior) attributes compared to conventionally grown alternatives. They tend to perceive such products as having particular intrinsic (quality and safety) characteristics (Vindigni *et al.*, 2002). However, the information about an organic product is asymmetric, as a consumer may not detect the presence or absence of organic characteristics even after purchase and use. Consumers may only know that the product is organic when they are informed (Giannakas, 2002). Consequently quality signals, such as product labels are needed to transform credence characteristics into search attributes, thereby enabling buyers to more clearly assess product quality (Yiridoe *et al.*, 2005). Organic products are therefore mostly identified based on the organic labels and/or organic logos attached. Several studies have found a positive relationship between consumer purchase decision and organic product labelling (see Ameseder *et al.*, 2008; Chang and Kinnucan, 1991; Mathios, 1998; Wessels *et al.*, 1999).

A key benefit of quality attributes of food products is human health (Caswell, 2000). This seems to be especially important for *organic* food (see, Goldman and Clancey, 1991; Schifferstein and Oude Ophuis, 1998). Regarding an in Austria conducted study from Nielsen (2007) health issues are the most important purchase criteria for organic food. Some experts see organic food as a form of 'soft functional food' (Haas, 2007: unpublished presentation). The willingness to pay a premium price can be seen as an investment in 'good health' (see Grossmann, 1972). This links the food quality attributes with consumer demand for organic food. Right after health issues environmental issues were mentioned in the study as being the main reasons for purchasing organic food products. This is confirmed by several studies in other countries (Estes *et al.*, 1994; Gregory, 2000). Concerning a study from Ernst & Young (2007) for the consumers of organic food increasingly pay attention to ethic and social aspects of the products they purchase. Due to this study environmentally friendly means to purchase products with low packaging and with *Austrian origin*. It is to be underlined, that ethical and social aspects include making *informed* consumer choices. This requires knowledge and awareness of competing products by the consumer. Because organic products are credence goods, knowledge and awareness are critical in the consumer purchase decision.

Especially the environmental and ethical issues as well as the mentioned importance of the origin in this study are of importance, when talking about international trade of organic food. Transport distances and different transport modes may result in a higher environmental impact. Ethical issues such as social standards in production or processing are different in third countries. Last but not least the origin of the products may influence the perceived quality of the product. As mentioned above not only values and attitudes or the knowledge and awareness about an organic product may influence the purchase decision. It's widely accepted that social and demographic variables influence the purchase decision as well (Baker and Crosbie, 1993; Fotopoulos and

Krystallis, 2002; Hay, 1989; Huang, 1996; Yiridoe *et al.*, 2005). Due to this rather complex situation, a conceptual framework adapted from Yiridoe *et al.* (2005: 196) was used. As shown in Figure 2 three factors are distinguished: product related factors, consumer related factors and exogenous factors.

Product related factors are differentiated into product characteristics and the perceived attributes of the consumer. Product characteristics can be noticed and experienced by the consumer and are for instance quality signals (taste) or communication tools (labels, certification). Perceived attributes otherwise cannot be detected by the consumer and therefore rely on information. The perceived product attributes are influenced by product characteristics as well as the knowledge and the awareness of the consumer regarding organic products and its values and attitudes, as the arrows in the figure show.

Consumer related factors imply social and demographic variables on the one hand and values & attitudes of the customer and their knowledge and awareness on the other hand, which are influencing each other.

Exogenous factors, such as the market environment, legal standards, or food quality and safety standards influence the purchase decision as well. Organic products are purchased because of their uniqueness *compared to conventional food* from a customer point of view. The market of conventional food therefore is of high importance for the organic sector as well. A widely accepted argument is that food scandals in the conventional sector favor the purchase decision for

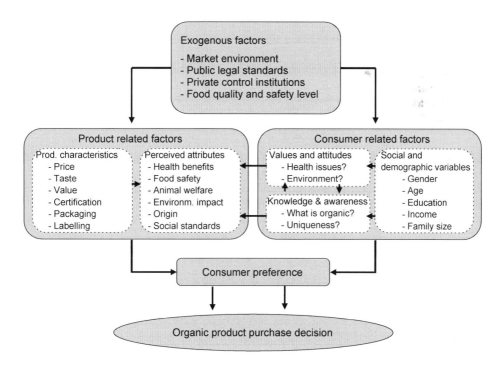

Figure 2. Framework of factors that affect purchase decision of consumers of organic food products (Yiridoe et al., *2005: 196).*

organic food (Yiridoe *et al.*, 2005). As well legal issues are of importance, such as the discussion of GMO food. All these exogenous aspects influence consumer related factors such as values, attitudes, knowledge and awareness and product related factors such as labelling or perception of food safety standards. Together the mutual influence of exogenous factors, product related and consumer related factors result in consumer preferences and furthermore in the purchase decision (see Figure 2).

When looking at the values and attitudes of consumers, environmental issues and social standards are of high importance, as mentioned before. As a consequence of global trade, transport distances grow constantly. Also with growing distance the probability increases that social standards in the country of origin differ strongly from standards in Europe or North America. Furthermore it has to be emphasized that for Austrian consumers organic food is highly linked to the Austrian origin (Nielsen, 2007). The relation between the organic origin of production and the credibility of an organic label is also verified by a study of the organic market in China. Wang *et al.* (1997) noted, that in regions of the world where the agricultural sector is not well developed, and the organic certification still is under way of standardization and harmonization, fewer consumers trust and believe in the organic label. So the credibility of an organic product seems highly connected to the origin of the organic production. One could say that the country of origin serves as an information cue for consumers about the development stage of the organic agriculture in a respective country, its professionalism in certification and its trustworthiness in the production and trading process. Due to the prevailing attitudes of European consumers we assume an *organic country of origin effect,* meaning that consumers use the attribute 'country of origin' as an information cue, which stands for:
- environmental friendliness/harmfulness due to different transport distances;
- higher/lower credibility of organic production methods;
- better/worse social standards.

Therefore it can be questioned, if the preference for an international traded organic product may be the same as the preference for a national produced organic one. Concerning our framework, the two factors – attitudes and knowledge & awareness – change in respect to international traded organic food. These two factors not only influence each other, but also influence the perceived attributes of an organic product (see Figure 2). This is of high importance, because as mentioned before consumers purchase organic products because they perceive them as unique or superior compared to conventionally grown alternatives. They tend to attribute to these products particularly intrinsic (quality and safety) characteristics (Vindigni *et al.*, 2002).

4. Legal issues on organic imports from third countries

Imports of organic products from third countries can only enter the EU under Article 11(6)[7], that is, with a special import permit issued by an individual EU member state, or relying on foreign certification services from importing countries (Xie *et al.*, 2005). In China for instance two main factors have prevented the implementation of an EU approved national certifications system: first, organic-based best farm and production practices are not commonly applied. In China lack of know-how, lack of storage and processing facilities, poor logistics and inadequate market information still exist (see Xie *et al.*, 2005). Second, within the EU organic plant products are legally protected under EU Regulation 1235/2008. This regulation was a step towards producer

---

[7] EU-General Food Law (Reg. 178/2002): 'Food and feed imported into the Community must comply with food law or conditions recognized as equivalent or, where a specific agreement exists, with requirements contained therein.'

and consumer protection within the EU. According to this regulation organic food imported into the EU by third countries must have been produced, processed and certified in accordance with equivalent standards. The exporting country must give details of the standard and inspection procedure implemented and these are evaluated by the EU. Yet foreign certification bodies entered the Chinese market (see Kledal *et al.*, 2007). Certification of organic export is done by internationally accredited companies like: OCIA (based in the United States), ECOCERT (French origin), BCS (based in Germany), IMO (based in Switzerland), Soil Association (based in the UK) and JONA (based in Japan) and by local certifiers such as Organic Food Development Centre (OFDC) or China Organic Food Certification Centre (for details about Chinese certification see Marchesini *et al.* (2010). According to Xie *et al.* (2005) this may push Chinese organic farming into different directions, as symbols and rules are different between different certification bodies. However, China's aim is to obtain equivalence to international standards, particularly standards from the EU, USA and Japan (Xie *et al.*, 2005).

5. Methodology

To determine the market potential of Chinese organic food in Austria, distribution channel operators and key players of the Austrian organic food sector were interviewed from November 2007 to January 2008. We assumed that the market orientation of the Austrian organic food chain is sufficient to communicate consumer preferences from one partner in the value chain to the next one. Grunert *et al.* (2003: 430) define 'market orientation of a value chain as chain members generation of intelligence pertaining to current and future end-user needs, dissemination of this intelligence across chain members, and chain wide responsiveness to it'. In a market oriented organic food value chain all are part of a system, which communicates consumer preferences from one partner in the value chain to the next one. Therefore qualitative interviews seemed appropriate to gather in depth information from business insiders and experts on all levels of the value chain, as purchase criteria of companies of different levels may vary. To cover the market as good as possible, different companies along the value chain were chosen: Bio Austria (i.e. the most important NGO of organic farmers in Austria), the two biggest wholesalers of organic food in Austria, representatives from Spar and REWE Austria (i.e. two retailers representing more than 50% of the Austrian retail market) two representatives from the first two organic supermarket chains in Austria and three partners from the organic food industry.

The analysis was inspired by methodologies developed by (Denzin and Lincoln, 2005). First the interview data were analyzed in an interpretive first order coding procedure, where text strings were coded describing purchase criteria of respondents especially for international purchased products. In a second step cross case analysis was carried out to identify the differences between companies and between companies of different levels in the value chain.

6. Results and discussion

Several factors were named that influence the purchase intention of channel operators and other key players of the branch. Interestingly, although these factors differ in detail, in general all interview partners more or less agree on the most important issues, when purchasing. First, interviewees named factors such as: product quality, price and localness/origin (in the sense of food with unambiguous regional provenance). Although often ranked in different orders, these factors are under the first issues named by the respondents. However, different channel distributors have different perception of product quality. For instance for conventional retail, it is very important that a product is available in a certain amount, and (mostly) for the whole year. Conventional retail faces the problem that it needs huge quantities to meet the demand of their customers. Therefore one of the most important criterion, when talking about product quality is, if the product is available in a certain quality and quantity. Furthermore a certain quality must be maintained the whole year or at least for a whole season (for seasonal products). Therefore in many cases small suppliers can hardly meet the needs of conventional retail. On the other side for managers of organic food stores and partly for Austrian wholesalers huge quantities are not of such importance. Therefore they can as well purchase regional products in small quantities.

One of the most interesting topics is the issue of localness. All interview partners agree that in Austria 'organic' is highly linked to the country or region of origin. Austrian consumers prefer Austrian products, and even more products related to a specific region in Austria. However, almost every interview partner had a different concept of how to define a regional product. While for managers and owners of specialized food stores a regional product is a product purchased in a region in its narrow sense, recognized as having a distinctive identity on the basis of its social and/or economic and/or natural characteristics. Given examples were regions from Austria such as Weinviertel (1,397 km^2) or Waldviertel (690 km^2). These are comparatively small distinctive regions coined by centuries of culturally specific patterns of producing and consuming food.

Managers from conventional supermarket chains see a regional product in a wider geographical sense, basically of Austrian origin. But retail chains located in Vienna also see products from neighbouring countries, such as Czech Republic, Slovakia or Hungary embraced by the term 'regional', as transport distances are shorter compared to longer transport routes to Western Austria (e.g. Salzburg, Tyrol). Wholesalers in general have the problem that their customers are basically organic food stores, and these are asking for regional products and regional brands in a narrow sense. But regional brands in Austria hardly exist; therefore wholesalers have to continuously find a trade-off between scarce regional products and 'abundant' international products. But for organic food stores the supply of distinct regional products is of strategic importance because they differentiate themselves from their competitors mainly by their assortment of regional products.

Coming to environmental issues, several interview partners mentioned that *organic food is rather a lifestyle philosophy* than just a product philosophy. From this point of view transport distances and the mode of transport are of interest for the whole sector. Almost all respondents try to avoid airfreight; only retailers still accept airfreight products for quality reasons. Due to transport distances respondents prefer to purchase in Europe or neighboring states. However, a second reason was mentioned: within Europe imports are easier than from third countries. Especially the food industry and wholesaler noted, that they regularly visit their suppliers. Establishing good relationships with their suppliers was mentioned as important for a number of respondents. Due to the cultural gap and the fact, that it's often not possible to visit suppliers from China, respondents preferably purchase in EU countries.

In consequence of these purchase criteria of Austrian Channel operators and the high preference for regional and Austrian products, market potential for Chinese products was evaluated as rather low. Only products are purchased on an international scale that are not available in Austria. Still, first of all, imports come from the European Union, mainly from neighboring countries. In the case of supply bottlenecks respondents rather try to establish good relationships with their suppliers, and work together on solutions. Often they accept that a specific product cannot be offered over the whole year. In some cases even the product recipe was changed, due to supply bottlenecks. Different strategies exist to deal with supply bottlenecks, purchasing in China or other Asian countries is only one of them.

However, almost all respondents stated, that sooner or later a number of products would be imported from China or other emerging countries. Especially discounters are expected to be the first offering a significant quantity of organic food from China. Till now, respondents could only name a few organic products already purchased from China, like garlic, pumpkin seeds for bakeries, ascorbic acid and citric acid.

It can be stated that, the preferences of channel distributors confirm widely with that of their consumers. Austrian origin, environmental issues as well as social aspects are of high importance for consumers of organic food items as well as for the key players of the sector. Shortage of supply was in most cases described as one of the biggest problems. Nevertheless, interviewed key players differentiate carefully between products, which can and should be purchased in Austria, and products, that are purchased internationally. Only products that correspond to their concept of an organic product and consequently to the values and attitudes of the consumers are purchased. For instance one of the interviewed retail chains started a project recently to help Austrian farmers increasing the local, i.e. regional, supply for fresh fruits and vegetables, because they are perceived to be very important, other product categories are of less importance. The interviewees emphasized on several occasions that – due to their regular personal contact to consumers and business partners – they are well aware of the prevailing consumer preferences. Thus the Austrian organic food value chain is a typical example for a value chain with high market orientation sensibly paying attention to the preferences of their consumers (Grunert *et al.*, 2005). Key players and channel operators therefore operate as much as gatekeepers, as they are deciders and influencers for the whole organic food chain.

Summarizing these results in respect to the initial research questions we can state that in the case of China there is a negative *organic country of origin effect*. Almost all respondents reported that they prefer to have personal relations with their business partners and if possible they prefer to visit their local production facilities. So, the rejection of China has mainly three dimensions. First, the channel operators argue that Austrian consumers prefer local, regional supply because environmental and social justice factors are an issue for them. As long as a product can be ordered regional, there is no need even in case of price differences to order internationally. Secondly, the channel operators themselves see long distance as a potential risk, because they cannot control the practices of their business partners. Third, the image of China as a reliable business partner, especially in the food sector, was questioned due to former food scandals. Insofar the interview partners confirmed that they connect trustworthiness and reliability directly with the image of a country, which is a clear country of origin effect. The products purchased in China can be seen as niche products. The interview partner see potential for future organic Chinese products if they are exotic and/or are connected to the image of traditional Chinese medicine such as pomegranate or ginseng.

Despite the conventional purchase criteria 'price', 'quality' and 'volume', the organic food value chain pays specific attention to environmental (transport distances), social and reliability issues (trustworthiness of certification and production itself). Localness serves as *the quality cue*, like an umbrella under which the consumer assumes that environmental, social and reliability standards are fulfilled. Nevertheless supply bottlenecks are one of the biggest problems for the channel operators.

These results pose further questions. Based on the life cycle analysis the organic market is somewhere between growth and maturity phase in Austria and growth phase in Europe (the decline in sales growth due to the economic crisis from 2008 is most probably a temporary effect). A maturing market for organic produce is reaching more and more 'conventional' consumers instead of the former 'core idealistic organic' consumers. Will these new and less idealistic consumers show the same preference for localness, using it as a quality cue for short distance transports and trustworthiness in the organic production? Or will these consumers pay less attention to the origin of the organic food over time, because organic as a quality cue per se is enough? Or will shortage in supply force the channel operators to 'educate' the consumers to accept more and more organic products from countries outside the EU? While growing markets show substantial profit improvement, profits stabilize or decline in maturing markets because of increased competition (see Kotler and Keller, 2008). This may be the reason why key players expect especially discounters to be the first to purchase Chinese organic food on a bigger scale due to their lower price.

Furthermore, studies in other European countries should investigate if the Austrian preference for localness is an exception or the rule compared to the preferences of other European consumers.

Discussion in lecture room

Apply your theoretical knowledge with respect to this paper:
- Discuss how far the opinions of the Austrian channel operators regarding Chinese organic food reflect the preferences of Austrian consumers? Read the article of Grunert *et al.*, (2005) *Market orientation of value chains* to learn about the theoretical background.
- Many Austrian consumers see organic products as *regional* products. Do you think this is only true for the Austrian market or is it a European wide phenomenon?
- Apply the life cycle theory on the development of the organic market. At which stage is your national organic food market? Compare it with other national organic food markets.
- Can the phenomena of food scandals or farmers selling conventional products as organic food be related to the stage at the life cycle? In other words do fast growing markets with unfulfilled demand attract a higher number of agents with opportunistic behavior?

Acknowledgements

The results reported in this paper derive from research activities promoted in the framework of the BEAN-QUORUM project (TH/Asialink/006), co-funded by the EU within the Europe-aid Asia-link Co-operation Programme.

References

AMA Marketing, 2007. RollAMA Motivanalyse-Bioprodukte. Available at: http://www.ama-marketing.at/home/groups/7/Konsumverhalten_Bio.pdf. Accessed January 2009.

Ameseder, C., Haas, R. and Meixner, O., 2008. Viability of values and attitudes concerning purchase intentions and benefit attribution for an organic sport drink. Conference paper for the 110th EAAE Seminar 'System Dynamics and Innovation in Food Networks, Innbruck-Igls, Austria.

Baker, G.A. and Crosbie, P.J., 1993. Measuring food safety preference: identifying consumersegments. Journal of Agricultural and Resource Economics 18: 277-287.

Brickner, I., 2007. Klimafeind Bio Erdbeere. 8 February, Der Standard, 8.

Caswell, J.A., 2000. Valuing the benefits and costs of improved food safety and nutrition. Australian Journal of Agricultural and Resource Economics 42: 409-424.

Chang, H and Kinnucan, H.W., 1991. Advertising, information and product quality: the case of butter. American Journal of Agricultural Economics 73: 1195-1203.

Denzin, N.K. and Lincoln, Y., 2005. The Sage handbook of qualitative research. 3rd edition, Sage Publications, New York, USA.

Ernst and Young, 2007. LOHAS - Lifestyle of Health and Sustainability. Available at: http://www.ama-marketing. at/home/groups/7/Konsumverhalten_Bio.pdf (12/2007). Accessed January 2009.

Estes, E.A., Herrera, J.E. and Bender, M., 1994. Organic produce sales within North Carolina: a survey of buyer options. Department of Agricultural and Resource Economics, North Carolina State University, Raleigh, NC, USA.

Fotopoulos, C. and Krystallis, A., 2002. Organic product avoidance: Reasons for rejection and potential buyers' identification in a countryside survey. British Food Journal 104: 233-260.

Giannakas, K., 2002. Information asymmetries and consumption decisions in organic food product markets. Canadian Journal of Agricultural Economics 50: 35-50.

Goldman, B.J. and Clancey, K.L., 1991. A survey of organic produce purchases and related attitudes of food cooperative shoppers. American Journal of Alternative Agriculture 6: 89-96.

Gregory, N.G., 2000. Consumer concerns about food. Outlook on Agriculture 29: 251-257.

Grossman, M., 1972. On the concept of health capital and the demand for health. Journal of Policy Economy 80: 223-255.

Grunert, K.G., Jeppesen, L.F., Jespersen, K.R., Sonne, A-M., Hansen, K., Trondsen, T. and Young, J.A., 2005. Market orientation of value chains. European Journal of Marketing 39: 428-455.

Haumann, B., 2007. Organic Farming in North America. In: H. Willer, M. Yussefi-Menzler and N. Sorensen (eds.), The world of Organic Agriculture – Statistics and Emerging Trends 2007, IFOAM Bonn, Germany, pp. 197-208.

Hay, J., 1989. The consumer perspective on organic food. Canadian Institute of Food Science Technology Journal 22: 95-99.

Huang, C.L., 1996. Consumer preference and attitudes toward organically grown produce. European Review of Agricultural Economics 23: 331-342.

Kledal, P.R., Hui, Y.Q., Egelyng H., Yunguan X., Halsberg N. and Xianjun L., 2007. Country report: Organic food and farming in China. In: H. Willer, M. Yussefi-Menzler and N. Sorensen (eds.), The world of Organic Agriculture – Statistics and Emerging Trends 2007, IFOAM Bonn, Germany, pp. 114-119.

Kotler, P. and Keller, K.L., 2008. Marketing Management. 13th edition, Prentice-Hall International, Upper Saddle River, New Jersey.

Marchesini, S., Huliyeti, H. and Spadoni, R., 2010. The market of organic and green food in China. In: R. Haas, M. Canavari, B. Slee, C. Tong and B. Anurugsa (eds.), Looking east looking west: organic and quality food marketing in Asia and Europe. Wageningen Academic Publishers, Wageningen, the Netherlands, pp. 155-171.

Mathios, A.D., 1998. The importance of nutrition labeling and health claim regulation on product choice: an analysis of the cooking oil market. Agricultural and Resource Economics Review 27: 159-168.

Nielsen, 2007. Bio Studie. Eine Nielsen Studie mit Analysen aus Nielsen Handelspanel und nationaler und internationaler Konsumentenforschung, Oktober 2007. Available at: at.nielsen.com/news/documents/ BioCharts_Presse.pdf. Accessed August 2009.

Padel, S., Jashinska, A., Rippin, M., Schaak, D., and Willer, J., 2008. The European Market of Organic Food in 2006. In: The world of Organic Agriculture – Statistics and Emerging Trends 2008; Edited by Willer H., Yussefi-Menzler M., Sorensen; IFOAM Bonn, Germany.

Richter, T. and Padel, S., 2007. The European Market for Organic food. In: H. Willer, M. Yussefi-Menzler and N. Sorensen (eds.), The world of Organic Agriculture – Statistics and Emerging Trends 2007; IFOAM Bonn, Germany, pp. 143-154.

Rippin, M., Kasbohrn A., Engelhardt, H., Schaak, D. and Hamm, U., 2007. Ökomarkt Jahrbuch 2007. Verkaufspreise im Ökologischen Landbau 2005/2006, ZMP, Bonn. Available at: http:// www.zmp.de/shop/mzm/oekomarkt_ jahrbuch_mzm_68.asp. Accessed August 2009.

Sahota, A., 2008. The Global Market of Food & Drink. In: H. Willer, M. Yussefi-Menzler and N. Sorensen (eds.), The world of Organic Agriculture – Statistics and Emerging Trends 2007; IFOAM Bonn, Germany.

Schifferstein, H.N.J. and Oude Ophuis, P.A.M., 1998. Health-related determinants of organic food consumption in the Netherlands. Food Quality and Preference 9: 119-133.

USDA, 2006. China, Peoples Republic of Organic Products and Agriculture in China 2006. USDA Foreign Agricultural Service, GAIN report number CH6405, 21 June 2006. Available at: http://www.fas.usda.gov/gainfiles/200606/146198045.pdf. Accessed August 2009.

Vindigni, G., Janssen, M.A. and Jager, W., 2002. Organic food consumption: A multi-theoretical framework of consumer decision-making. British Food Journal 104: 624-642.

Vogl, C.R., Kummer, S. and Haas, R., 2005. Organic Farming in Austria – Idealism vs. Market realism in the organic movement. Case Study developed for the Global Seminar. Available at: www.globalseminar.org. Accessed August 2009.

Wang, Q., Halbrendt, C. and Webb, S., 1997. Consumer demand for organic food in China: Evidence from survey data. In: W. Lockretz (ed.), Agricultural Production and Nutrition. Conference Proceedings Boston, MA, 19-21 march. Tufts University Boston, MA, USA, pp. 187-194.

Wessels, C.R., Johnston, R.J. and Donath, H., 1999. Assessing consumer preference for eco labeled seafood: The influence of species, certifier and household attributes. American Journal of Agricultural Economics 81: 1084-1089.

Willer, H., Sorenson, N. and Yuseffi-Melzer, M., 2008. The world of Organic Agriculture 2008: Summary. In: H. Willer, M. Yussefi-Menzler (eds.), The world of Organic Agriculture – Statistics and Emerging Trends 2008. IFOAM Bonn, Germany.

Willer, H. and Yussefi-Menzler, M., 2008. The world of Organic Agriculture – Statistics and Emerging Trends 2008; IFOAM Bonn, Germany.

Xie, B., Li, T., Zhao, K., and Xi, Y., 2005. Impact of the EU organic certification regulation on organic exports from China. Outlook on Agriculture 34: 141-147.

Yiridoe, E.K., Bonthi-Ankomah, S. and Martin, R.C., 2005. Comparison of consumer perception and preferences towards organic versus conventionally produced foods: A review ans update of the literature. Renewable Agriculture and Food Systems 20: 193-205.

Yussefi, M. and Willer, H., 2007. Organic Farming Worldwide 2007: Overview & Main Statistiks. In: H. Willer and M. Yussefi (eds.), The world of Organic Agriculture – Statistics and Emerging Trends 2007. IFOAM Bonn: Eigenverlag.

Marketing high quality Thai organic products in Europe? An exploratory approach

P. Lombardi[1], M. Canavari[1], R. Spadoni[1], R. Wongprawmas[2], B. Slee[3], D. Feliciano[3], B. Riedel[4], M. Papadopoulou[1] and F. Marin[5]
[1]*Alma Mater Studiorum University of Bologna (UniBO), Department of Agricultural Economics and Engineering Viale Fanin 50, 40127 Bologna, Italy*
[2]*Thammasat University, Faculty of Science and Technology, Thammasat University Rangsit Center, 99 Paholyothin Road , Klongluang, Patumtani 12121, Thailand*
[3]*Macaulay Institute, Craigiebuckler, AB15 8QH Aberdeen, United Kingdom*
[4]*Humboldt University, Unter den Linden 6, 10099 Berlin, Germany*
[5]*University of Trento, via Belenzani, 12, 38122 Trento, Italy*

Abstract

In an international trade context, environmental elements (e.g. the introduction of organic agriculture) and the role of the country of origin could influence the demand in the market of destination. This study aims at describing how qualitative research methods may help deepen knowledge regarding interest towards organic products imported from Thailand, specifically Kamut® wheat, organic rice and tapioca. In the case of Kamut-based-products 21 individual interviews were conducted with Italian large scale retailers, whereas in the case of products based on Thai rice and tapioca, 4 focus groups comprising of European consumers from Germany, Greece, Italy, and Scotland were administered. The individual and group interviews were recorded, transcribed and analysed using a qualitative approach. In both cases a list of semantic categories was created, explained and supported by analysing the discussions using content analysis techniques. According to these results, survey participants did not know much about these products, especially about tapioca. In general, Thai Organic rice was perceived by the interviewed consumers as a 'different type' of rice and tapioca as a 'new food product'. Kamut® wheat was also perceived by the interviewed retailers as a 'new product'. The survey participants tend to favour the product's nutritional aspects, as the most relevant attributes followed by taste and smell, as well as a series of social and environmental benefits. The information obtained could be useful in further exploration of this topic, but it needs to be tested with a quantitative approach.

1. Introduction

The differentiating elements, that can influence trade interest in the international context and allow a commodity to be converted into a differentiated food product belong to several categories, such as the environmental aspects (e.g. application of production methods that better safeguard the environment), the origin/production country (e.g. the interest towards food products coming from countries where the food habits and production methods are very different) or the nutritional-health aspects (potential substitutes for the usual food products in case of intolerances and/or allergies), etc.

In this survey, differentiated products imported by Thailand are considered: Kamut® wheat, organic rice and tapioca. They are all organic products. The European Union (EU) is one of the most important potential destination markets; additionally some aspects of these imported products can have a strong impact on the food habits of the importing countries. The main objective of this survey is twofold:

1. to explore the elements that could influence the interest of Italian large scale retailers (in the Kamut-based products case); and
2. to analyse European consumer attitudes (in the case of rice and tapioca based-products).

The research approach adopted is designed to give useful information to researchers and practitioners in the food sector who are interested in acquiring better awareness of the development of a specific EU-Thailand trade channel. However, it is not intended to provide quantitative information on the issues examined.

These kind of issues are instead suited to qualitative research techniques, because they allow the generation of new ideas and selection of a limited set of possible alternative solutions for new or not well known products/services. The chosen qualitative research technique allows delimitation of the survey objective and generation of a hypothesis to be eventually tested with further quantitative study (Molteni and Troilo, 2003). The qualitative elements emerged from the data, could however, already support the development of an information base with an ability to improve the knowledge of the characteristics and versatility of these products.

In the analysis of Kamut-based products, face-to-face in-depth interviews with practitioners working in the large retail industry were conducted. This data collection method was deemed more suitable than interviewing the Italian large retail managers, because of their position in competing companies and their limited availability of time.

In the case of organic rice and tapioca-based products, focus groups were conducted since this technique is more adequate to rapidly provide qualitative information from a wide geographical scope (the European context) and it is much simpler to gather 6-8 consumers and organise a group interview than arrange individual interviews.

The first section of this paper involves analysis of the market situation in the European and Italian context to understand the background conditions for a possible interest towards rice, tapioca and Kamut® wheat; then, the general objectives of the two analyses are described; in the following sections a detailed description of the methodology and the main results are presented; finally, we propose some comments about the effectiveness of the research tools adopted.

2. Background

2.1 Organic agriculture in Thailand

Organic agriculture is still in its early stage in Thailand. However, in the last years an increasing interest towards organic products of tropical origin is emerging in developed countries (Green Net/Earth Net Foundation, 2008). According to previous studies (Kung Wai, 2004) it seems that exporting opportunities are the major driver for the conversion to organic agriculture in all Asian countries, including Thailand.

Most of the organic products sold are basic unprocessed commodities such as rice, fresh fruits and vegetables. Increasingly, more intermediate processed products are developed, such as sugar, tapioca starch and palm oil (Panyakul, 2003).

2.2 Kamut®

Kamut® is a registered trademark and a brand associated with a special durum wheat variety (*Triticum turgidum* spp. Turanicum, also known as Khorasan wheat), grown using the prescriptions of organic agriculture. Kamut® is genealogically similar to modern durum wheat. According to the information provided by the brand owners, it has been rediscovered by a US farmer (http://www.kamut.com) after a long period of obscurity. Currently a growing interest for the product has developed in the market category of high-quality food products.

This product is differentiated from regular wheat because of its intrinsic characteristics including higher nutritional value, particular taste and freshness, high digestibility, high content in selenium and extrinsic characteristics (particularly the fact that it is organically grown and certified) allowing the brand owner to adopt a value-enhancement marketing policy. The brand owners claim that in ancient times it was grown in the area between Egypt and Mesopotamia, but nowadays its cultivation spreads mainly throughout North-America. Currently, it is also experimentally grown in different areas of the world and field tests for this crop are performed in Thailand, which is considered the most promising area for further development of the production areas.

Europe, represents the most important destination market for Kamut® wheat. Italy is the largest EU consumer market for Kamut® and imports approximately 70% of all the Kamut® wheat exported to Europe. Currently, Italy is the European point of reference for the production and innovation regarding Kamut-based products.

Most of the Italian Kamut-based products are processed and re-exported into other EU countries. However, in the second half of 2006, a growing share of domestic demand was reported. Domestic marketing efforts shifted from a predominant distribution to specialized food stores (e.g. specialty health stores, herbal medicine stores, and organic food stores) to the distribution of Kamut-based products to large-scale retail chains. Therefore, a preliminary analysis of the marketing and distribution channel in Italy may provide fundamental information to support the strategic marketing decision regarding the viability of a refocusing of Kamut® distribution in Italy.

2.3 Rice and tapioca

Rice and tapioca are considered two of the most important and peculiar food products of Thailand (Roitner-Schobesberger *et al.*, 2008). From the trade point of view, the market structure for these commodities is driven by a system of import quotas. The marketing opportunities are limited by the achievement of these quotas and/or, in the future, by the possible re-negotiation of international agreements.

Although Thailand is not the leading producer of rice, it exports more rice than any other country in the world, accounting for 29% of all rice exported (Thai rice Exporter's Association: http://www.thairiceexporters.or.th). Rice (both white and brown rice such as Kao Hom Mali, Lueang-On, Red Hom Mali) is also the most important organic crop in Thailand. The bulk of Thai organic produce is exported to the EU, with the remainder destined mainly to Japan, the US and Singapore. The major organic product approved for export to the EU is jasmine rice (Suwannaporn and Linnemann, 2008), which can be certified by a variety of certification bodies, i.e. Bioagricert, KRAV, ACT, BCS, The Soil Association and Ecocert.

Tapioca is the starch extracted from the root of the plant species *Manihot esculenta* (Crantz), known as Manioc or Cassava, which is the third most important crop in Thailand (Office of

Agriculture Economics of Thailand: http://www.oae.go.th). Thailand is the world's largest producer and exporter of tapioca starch and starch derivatives in the world, with an annual production of over 2 million tons of starch (Department of Foreign Trade of Thailand: http://dft.moc.go.th). The European Union (EU) is the most important destination for Thai tapioca, but its main use is as an ingredient in the feed industry (The Thai tapioca Trade Association: http://www.tta-tapioca.org/products.php), therefore it can be considered a typical commodity. In this study, however, we focus on tapioca as a specialty organic product that can be found in supermarkets and used as a flour ingredient in special exotic recipes and as a high digestible thickener in baby-food.

3. Objectives

In the study regarding Kamut-based products, the main objective was the exploration of potential opportunities for these products in Italian large-scale retail chains. The study is further specified through five sub-objectives, as follows:
1. to explore the main reasons that might induce distribution operators to introduce Kamut-based products to their shelves;
2. to evaluate the expectations of distribution operators about potential reactions of consumers;
3. to identify the limitations regarding the introduction of Kamut-based products to the shelves;
4. to specify the relevant characteristics regarding consumers and retailers' quality perception;
5. to list the potential competitors for Kamut-based products' market share.

The evaluation of the interest of the European consumers towards Thai organic rice and tapioca was the objective of the second study. Also in this case specific sub-objectives were explored, these included:
- the set of characteristics (attributes) that may be considered the most relevant for the perception of the product quality by the consumer;
- the possible factors that may be able to influence positively the consumers interest;
- the possible factors that may be able to influence negatively the consumers interest;
- reasonable price brackets for European consumers;
- trust elements (including the EU certification schemes).

4. Methodology

When research is conducted using an explorative approach, qualitative analysis techniques are often the most suitable choice, since in this way the researcher is able to grasp the complexity of phenomena, situations and events. Thus, allowing the revelation of variables, which represent such occurrences and the ability to figure out the relationships between the detected variables.

Qualitative analysis techniques are generally used to acquire an 'in-depth' knowledge, which highlights aspects that are at times hidden but that may become significant in the future. They as well formulate research hypotheses that may then be tested using a quantitative research approach, when necessary. Furthermore, in qualitative techniques only few subjects are usually involved.

The most frequently used techniques employed for the collection of qualitative data are focus groups (where the participants usually have to discuss issues together, with the help of a moderator who monitors the discussion) and face-to-face individual interviews, which allow a direct relationship between the interviewee and the interviewer.

In a qualitative study, sample representativeness is usually not a stringent requisite, and in any case it cannot be achieved using random sampling. Therefore, the participants in a qualitative survey are normally recruited according to criteria of opportunity and convenience, amongst individuals who are deemed able to provide information useful to the investigation, meanwhile trying to ensure a balanced representation of different types of participants. In both group and individual interviews, the interviewer's task is to collate information, taking care not to influence the interviewee. This is important to remain as objective as possible when processing and interpreting responses. For a discussion on the features of different qualitative research techniques see Aaker *et al.* (2007).

4.1 Direct in-depth interviewing technique: Kamut® case study

A direct in-depth interview can be defined as 'an unstructured personal interview which uses extensive probing to get a single respondent to talk freely and to express detailed beliefs and feelings on a topic' (Webb, 1995).

For the purposes of qualitative analysis, it is not necessary to conduct a large number of interviews, but it is essential to select the appropriate subjects: interviewees who are able to adequately highlight the most relevant problems for the objectives identified. The depth and breadth of the interview are also important.

The presence of an interviewer and an assistant is generally required to conduct this type of interview. The assistant has the task of taking note of the main conversation elements and to record the conversation (if authorized by the interviewee). The methodological process used in the Kamut® study entailed the following steps:

- assimilation of 'topics' and 'inputs' by the interviewer, with the help of a pre-defined interview outline;
- preparation of informative materials and recording equipment;
- administration of the interview;
- preparation of an interview summary aimed at highlighting those elements deemed most important to an initial analysis;
- transcription of the recorded conversation (if authorized);
- analysis of the whole documentation.

This research procedure is particularly flexible and it allows the problems raised to be discussed freely and openly in terms of possible interpretations by the interviewees. Whilst this issue represents an important advantage, it must nevertheless be underlined that this method of survey also presents notable disadvantages which are inherent in the technique and difficult to overcome. These weaknesses involve the following: the method of administration is not standardised, there is the significant possibility that responses are influenced by the interviewer's attitude and mode of questioning, results are not easy to interpret and may entail a certain degree of subjectivity. Nevertheless for the purposes of this study, the advantages of this method are deemed greater than the possible disadvantages.

In order to define the reference sample, a non-probabilistic sampling method was used for the selection of the subjects to be interviewed. In this case, an attempt was made to ensure a complete coverage of the Italian market with respect to the distribution of outlets and/or type of distribution organisation (co-operatives, private companies, associated/leader companies, smaller businesses).

In total, 43 companies were selected. For each company, buyers and marketing managers were identified as potential interviewees and contacted by phone. All of the selected individuals were invited to participate in the survey asking them to accept an interview in person with a researcher or to fix an appointment for a telephone interview. Overall, 21 interviews were conducted (11 in person at the distributor's workplace and 10 by phone). The interview duration lasted between 45 and 60 minutes.

A semi-structured interview outline was used as support to the discussion. It was composed of a series of discussion points (objectives); alongside each objective a series of possible 'inputs' was listed, consisting of specific questions or statements which the interviewer could use as a tool to encourage interviewees to express their opinions. The purpose of the outline was to provide a guide to the interviewer, in order to assure that the conversation covered all of the relevant objectives. However, the interviewees had the possibility to freely express their opinions.

4.2 Focus group technique: rice and tapioca case study

For the survey on the possible elements of interest for European consumers towards Thai organic rice and tapioca, the focus group technique was used. A focus group can be defined as: '.....a group of individuals selected and assembled by researchers to discuss and comment upon, from personal experience, the topic that is the subject of the research' (Gibbs, 1997).

Every survey technique presents limits and resources in relation to the objective of the research. Despite this, Zikmund (1997) summarised the advantages of such focus group in ten elements defined the '10 Ss', which are subsequently mentioned by Stokes *et al.* (2006).

It would appear that the most important benefits of focus groups derive from two features: group interaction (Albrecht *et al.*, 1993; Burns, 1989) and the replication of social forces (Krueger and Casey, 2000; Robson, 1990). It is clear that the process of group dynamics, responsible for many of the advantages of focus groups, can also be regarded as a disadvantage. In the case of research, which has the objective to evaluate private aspects, situations involving public interaction may inhibit participant involvement (Greenbaum, 1998; Hedges, 1985).

Some disadvantages may also be detected in the following situations: when the moderator cannot maintain his role and the interaction between the participants is therefore disturbed by his presence; even if the focus group allows the collection of much information in a brief timeframe, it requires considerable resources (costs and management aspects); if the moderator is not adequately trained, the discussion could become uniform and therefore inadequate to provide a useful input for the subsequent analysis.

Advantages and disadvantages mentioned above are mediated by the specific context and situation and can only give us an orientation about the usefulness and limitations of focus group interviews. However its adoption must be considered when comparing it with other research techniques.

Focus groups are particularly adequate to understand the behaviours regarding consumer choice (including food choice), because it encourages participants to express their opinions and to confront each other (Morgan and Krueger, 1993). Moreover, focus groups allow us to explore consumer preferences, consumer motivations and limits connected with these preferences. The source of information is the 'group' and not a single person; the most important element is the interaction between participants, not only the individual's opinion.

Normally, during the discussion, there are both a moderator, whose task is to conduct the discussions between the participants, and one or more assistants who have to manage secondary activities. The moderator can be totally absent from the discussion (it is the point of view of the observer) or they can be present, but with a limited degree of intervention. They should not express their own opinions about the discussion, but can introduce 'doubts', 'taunts', 'input', and 'themes' in order to stimulate it (Zammuner, 2003). The participants should try to avoid bringing negative contributions into the discussion and to control their reaction to disagreeing opinions.

Usually, the focus group structure differs on the basis of three specific parameters: the composition of the group (more or less homogeneous), the degree of structure (from totally de-structured focus group, where there are no pre-defined questions to a totally structured focus group, where every intervention responds to a specific input) and the role of the moderator (absent, moderate, extreme).

In this study about Thai organic rice and tapioca, four focus groups were conducted in different European countries, specifically in Germany, Greece, Italy and Scotland (UK). The groups were each composed of 6-8 persons and the average duration of each discussion was from 60 to 90 minutes.

The participants (altogether 28) were recruited using a convenience sampling method on the basis of three main characteristics: purchasing at the points of sales of large scale retail distribution chains (usually), purchasing organic products (at least sometimes), consumption of rice or rice-based products (at least sometimes).

Initially, every moderator was conveniently trained about the planning of the discussion, the recruitment of the participants, the conduction of the discussion and the elaboration of collected data. During the discussion, every moderator was supported by a semi-structured interview outline which has the same function described for the direct in-depth interview. Finally, every moderator managed and interpreted the collected data.

4.3 Content analysis

Processing of collated qualitative data entails the use of techniques which involve a varying degree of subjectivity as well as analytical tools, which are intuitive or supported by data processing software. A qualitative approach to content analysis consists of the (subjective) interpretation of the content of text data, through the systematic classification process of coding and identifying themes or patterns (Hsieh and Shannon, 2005). The essential objective of our qualitative content analysis, performed as a heuristic analysis of the textual content of the transcripts, is to present the most interesting elements arising from each discussion and/or conversation in order to gain an extensive overview of interviewees' attitudes towards the themes under investigation.

At the end of the interviews/discussions, the assistants immediately summarised the opinion of the participants, the aspects of the non-oral communication and para-linguistics aspects (gestures, postures, positions). Each interviewer listened quickly to the recorded conversation. Subsequently, they transcribed the whole conversation: in the case of the focus groups transcription required more time and attention than the direct interviews because of the overlapping of different voices.

Finally, a conventional heuristic content analysis approach was adopted to analyse the transcriptions of recorded discussions. Conventional content analysis consists of a study design where the aim is to describe a phenomenon, avoiding the use of preconceived categories, instead allowing the categories to emerge from the data, through an inductive category development process (Hsieh and Shannon, 2005; Kondracki and Wellman, 2002; Mayring, 2000). This analysis entailed the following steps:

- to read the transcriptions carefully;
- to highlight the sub-texts of the transcription that corresponded to the research specific objectives;
- to create a system of codes that represent different 'semantic categories';
- to assign the different interventions to the same code whereas they represent the same semantic category (Bloor *et al.*, 2002);
- to give a synthetic and neutral name to every semantic category (without negative or positive characterisation);
- to explain the meaning of each category with sentences extrapolated from the transcriptions (that can often have opposite semantic content) and brief comments (Krippendorf, 1983).

5. Results and discussion

In this section for the sake of brevity, we will not discuss all of the semantic categories identified in the content analysis of the text data derived from the interviews and the focus groups. As examples of the analysis conducted in the two case studies, we only present a shortened version of the analysis of one semantic category for both the Kamut-based and rice/tapioca-based product cases. This does not imply that these two categories are deemed more important than the cases not discussed in the following sections.

5.1 An example of semantic category analysis in the Kamut® case study (Italian large scale retail structure)

The problems related to logistics-management issues are prominent and mainly concern the frequent delivery of a wide product quantity and inventory turnover. Unfortunately, these issues include the limited diffusion of Kamut-based products, the fact that these products are not always available (due to limited cultivation areas), and current low consumer demand that limit high turnover on the shelves. Additionally, the evolution of Italian large-scale retail businesses leads to a major presence of medium-small size structures (supermarkets and minimarkets), where limited shelf capacity represents a substantial obstruction for the introduction of new and unfamiliar products: '...the real nature of large retail distribution means the product must be rotated and therefore must guarantee a certain type of rotation and it is also true that large retail distribution is also full of normal, basic products which aren't actually rotated, it then depends to some extent how this product also builds up a culture in the large retail distribution sector...'.

The Italian food processing industry has a highly fragmented structure; it lacks product range both in terms of width and depth and can hardly promote niche Kamut production. For this reason, the Kamut® introduction should be obtained through food supply chains managed by a national intermediate agent who can correctly manage the supply: '...Today's fragmented situation cannot build culture...the presence of a sizeable point of contact also creates more clarity in the market, as well as in the offer, since there is no offer at the moment...'.

5.2. An example of semantic category analysis in the rice and tapioca case study (environmental aspects: agricultural production method)

Participants' opinions are mainly in favour of organic agriculture which, without use of pesticides, aims to protect farmers' health, consumers' health and safeguard environmental resources: 'surely,

I'm interested in the agricultural production method with which the product I buy is produced; yes, if it's organic farming'.

Moreover, organic production is seen as a way which guarantees for the consumer, not only the environmental safeguard but also a better taste and healthiness. Despite this, one participant disagreed with this general opinion and she said that the organic agriculture attribute is a secondary aspect to purchasing Thai organic rice and tapioca. Aiming exclusively at the organic production method can not be a strategy of valorisation: 'I don't look for organic rice, because in general I don't buy organic products'.

In some participants' opinions, the concept or organic production is conflated with 'local' and the notion of imported organic food, particularly that imported from a great distance, generates dissonance with such consumers (pertaining to the 'food miles concept', Lang, 2006). 'Organic for me is more 'local', produced in organic way and bought in local markets rather than coming from China, or Chinese organic rice.

5.3 Semantic categories identified in the two studies

At the end of the qualitative content analysis conducted on the available text data, similarly to what we described in the previous sections, a list of semantic categories have been defined in correspondence to each specific objective. The complete list of the categories highlighted in both case studies is reported in Table 1 and Table 2.

In order to make appropriate use of the information provided by an explorative study based on qualitative techniques, the most important caution is to avoid inferences to generally valid conclusions starting from the results of such a study. However, some elements could be considered by the operators of this food sector to identify possible scenarios to foster the development of international trade of these products as differentiated goods.

It was apparent that participants still do not know much about these products. In general, Thai organic rice is perceived by the interviewed consumers as a 'different type' of rice, while tapioca is perceived as a 'new product'. Regarding Thai rice, it would be necessary to plan a promotion activity to value-enhance and better communicate the differentiating elements. Many doubts are cited about the nutritional, agronomics and use properties of tapioca.

Kamut® is generally perceived by the retail practitioners as a new product. There is much curiosity on nutritional and sensory aspects, while there are many doubts on economic and agronomics aspects. This initial situation, however, does not exclude the possibility to find encouraging trade and marketing solutions in order to introduce these products in the European market, since several possible strengths in comparison to existing products have been identified.

Table 1. List of semantic categories of the heuristic content analysis about Kamut-based products (retailers' opinions) (Own elaboration of survey data).

Reason for introduction and shelf positioning strategies
- Satisfaction of expectations' of specific target consumers
- Innovative product
- Expanding and adding depth to range offered
- Brand differentiation and enhancement
- Adaptability/transversal/versatility
- Drive for suppliers
- Nutritional/health aspects
- Functional food
- Pharmaceutical food
- Organic method production

Limiting factors regarding the introduction
- General knowledge about products
- Information/communication
- Price
- Food suppliers/food industry structure
- Italian large scale retailers structure
- Connection with culinary/cultural/geographic traditions of the trade importers
- Country of origin/production
- Product brand
- National processing brand
- Sensory aspects

Expectations of distribution operators regarding the consumers' reactions
- Interest due to specific need (obliged consumer)
- Interest due to product's qualitative characteristics (careful consumer)
- Curiosity
- Indifference (indifferent consumer)
- Need to differentiate food habits (experimental consumer)

Relevant attributes for the quality perception product
- Nutritional/health aspects
- Economic aspects
- Environmental aspects
- Hedonistic aspects
- Historical aspects (influence of country of origin)
- Packaging
- Perception of added value/marginality
- Availability of different type of Kamut-based products

Potential competitors
- Spelt-based products
- Rice and corn-based products
- Traditional wheat-based products

Table 2. List of semantic categories of the heuristic content analysis about rice and Tapioca-based products (consumers' opinions) (Own elaboration of survey data).

Relevant attributes for the quality product perception
- Health/nutritional aspects
- Environmental aspects: agricultural production methods
- Versatility
- Sensory aspects
- Aesthetic aspects (colour, structure, shape)
- Ethics aspects
- Hygienic-sanitary aspects

Factors which can positively influence the consumers' interest
- Research of food diversification/innovation
- Curiosity
- Attention to a specific physical need
- Research of a versatile product
- Family's influence
- Life style
- Marketing activities: promotion campaign, position in supermarket shelves, tasting in the supermarket
- Marketing activities in specialized Thai shops
- Packaging
- Attention to safeguard environmental aspects
- Attention to farmers' respect

Factors which can influence negatively the consumers' interest
- Connection to local food habits
- General knowledge/information
- Impact of increasing of numerous ethnic groups in the European countries
- Process certification driven by foreign countries
- Price
- Evolution of purchasing modality
- Country of origin
- Limited diffusion in the supermarket of Italian retail distribution chains

Proposal of reasonable price brackets
- An increase of willingness to pay
- No increase of willingness to pay
- Availability of average consumer

Trust elements
- Brand of distributor: private label
- Indication of country of origin
- EU organic certification production method
- Nutritional label
- Brand of producer
- Brand of European certification bodies
- Fair trade certification
- Advertising

6. Final remarks

The explorative approach and the type of qualitative techniques used in this work allowed us to extrapolate a series of ideas which could represent a support to the commercialisation of these products imported by Thailand. Evaluations by distributors and consumers can be very important for the producing countries to support future export activities.

In the Kamut® case study, opinions, beliefs, and attitudes among the large-scale retailers, help to illustrate the scenario regarding the trade of Kamut-based products. Some managerial implications may be derived from the analysis.

The research allowed us to differentiate the highlighted elements between motivating ones and limiting ones. The great versatility of Kamut® as a raw material, is that it is suitable for several consumer uses. Kamut® is organically grown, features numerous important quality attributes and is managed accordingly with a global value-enhancement strategy, made possible by its protected brand name.

This explorative analysis revealed that interesting opportunities exist in the Italian large-scale retail channel for a wide range of Kamut-based products. Probably, the commercialization of the product as a raw material could intensify trade activities between countries, considering that Italy has a strong processing tradition. However, the principal limit of this strategy is represented by the competition of other countries that could produce at lower prices. Perhaps in the European panorama, the opportunities for processed products would be more interesting.

According to recent statistical surveys, the solutions for the application of a trade strategy must put more emphasis on nutritional characteristics that meet the needs of a growing number of consumers. Moreover, communicating the historical/environmental/cultural features of Kamut® may satisfy the demand of people who have a vested interest in food culture and the effects that the organic farming method can have on the environment.

Summarising the rice and tapioca case study, the most important critical issues affecting participant's opinions, include the lack of trust in the certification process by foreign countries and an attitude against trying food novelties, associated with a sense of loyalty to (or affection for) local food traditions. The most important trust elements are represented by the brand of the distributor, the producer and the EU and national certification bodies; in particular, participants associate some parameters of guarantee and safety to the brand name.

In conclusion, this study has highlighted the important issues to better match the features of the supply from specific production regions (i.e. Thailand) to the needs, wants, and demands of the most important consumption countries (i.e. EU and Italy, respectively). The methodology used belongs to the domain of qualitative research, which typically entails the in-depth examination of a topic, but not allowing for substantial quantitative information. This qualitative research does not give results that could be applied to the market as a whole. Therefore, the information should only be used as suggestions to design a strategy, or maybe the starting point to generate hypotheses for subsequently describing the issues raised in the study. That is, the elements that emerged could be used to arrange a subsequent analysis based on a quantitative research approach, in order to more accurately define possible distributive strategies and demand requirements regarding the distribution channel for Kamut® and possible consumer strategies for rice and tapioca.

Discussion in lecture room

Apply your theoretical knowledge with respect to this paper:

- Formulate a specific marketing plan based on the results of this qualitative study, to increase exports to Italy for one of the investigated products (Thai rice, Kamut® or tapioca).
- Under the premise of a limited budget for communication purposes, which promotional activities would you foresee to increase knowledge of Italian consumers about Thai rice/Kamut®/tapioca? Search the marketing literature for 'below the line' promotional activities. Which of them seem to be the most useful?
- Discuss the advantages and disadvantages of *qualitative* marketing research methods compared to *quantitative* marketing research! Under which circumstances would you prefer to use a qualitative and/or a quantitative approach? When would you use focus groups instead of qualitative expert interviews?
- Look at the examples in the semantic category analysis on rice and tapioca! Compare these examples with the results of the study of Haas *et al.* (2010) about the potential of Chinese organic food products in Austria.

Acknowledgments

This paper is based on research activities jointly designed by Maurizio Canavari (Sections 3 and 4), Pamela Lombardi (Sections 5 and 6), and Roberta Spadoni (Sections 1 and 2). This research was partly funded by the Chang Mai rice Research Centre, Sanpatong, Chiang Mai 50120 (Thailand), in the framework of the project 'Small projects facility – growing organic kamut wheat in northern thailand for EU market' TH/SPF/G/02 (105754), and by the Subcontract to contract No. TH/SPF/G/11 (112318) for the EU-Project 'Small projects facility – enhancing competitiveness of organic rice and tapioca cultivations to stimulate local development and Thai-EU trade relations' granted to the Thammasat University, Bangkok, Thailand. Floriana Marin contributed performing some of the individual interviews. Bill Slee, Diana Feliciano, Maria Papadopoulou and Bettina Riedel, contributed performing the focus group interviews in Scotland, in Greece and in Germany. Rungsaran Wongprawmas collaborated in the interviews design and participated as observer to the focus group in Italy.

References

Aaker, D.A., Kumar, V. and Day, G.S., 2006. Marketing Research, 9th ed.: John Wiley & Sons, Hoboken, NJ, USA.

Albrecht T.L., Johnson G.M. and Walther, J.B., 1993. Understanding communications processes in focus group. In: D.L. Morgan (ed.), Successful Focus group: Advancing the state of art, Sage, Newbury Park, CA, USA, pp. 51-64.

Bloor M., Frankland J., Thomas M., and Robson K., 2002. Focus Groups in Social Research. Londra, Sage, Volume 3, No. 4.

Burns C., 1989. Individual interviews. In: S. Robson and A. Foster (eds.), Qualitative Research in Action. Hodder and Stoughton, Londra, pp. 47-57.

Gibbs A., 1997. Focus group. Social Research Update, Vol 19, Department of Sociology, University of Surrey, available at: www.soc.surrey.ac.uk/sru/sru19.html. Accessed August 2009.

Greenbaum, T.L., 1998. The Handbook for Focus Group Research. Sage, Thousand Oaks, CA, USA.

Green Net/Earth Net Foundation, 2008. Organic Statistics Thailand. Available at: www.greennet.or.th. Accessed August 2009.

Haas, R., Ameseder, C. and Liu, R., 2010. Factors influencing purchasing decisions of Austrian distribution channel operators towards 'made in China' organic foods. In: R. Haas, M., Canavari, B. Slee, C. Tong and B. Anurugsa (eds.), Looking east looking west: organic and quality food marketing in Asia and Europe. Wageningen Academic Publishers, Wageningen, the Netherlands, pp. 115-126.

Hedges A., 1985. Group interviewing. In: R.L. Walker (ed.), Applied Qualitative Research. Gower Publishing, Aldershot, UK, pp: 71-91.

Hsieh, H.-F. and Shannon, S.E., 2005. Three approaches to qualitative content analysis. Qualitative Health Research 15:1277-1288.

Krippendorf, K., 1983. Analisi del contenuto: Introduzione metodologica. ERI - Edizioni Rai radiotelevisione italiana, Torino, Italy.

Kondracki, N.L. and Wellman, N.S., 2002. Content analysis: Review of methods and their applications in nutrition education. Journal of Nutrition Education and Behavior 34: 224-230.

Kung Wai, O., 2004. Asia. In: H. Willer and M. Yussefi (eds.), The world of organic agriculture 2004. Statistics and emerging trends. IFOAM: Bonn, Germany, pp. 69-78.

Krueger, R.A. and Casey, M.A., 2000. Focus Groups: A practical guide for applied research. Sage, Thousand Oaks, CA, USA.

Lang, T., 2006. Locale/global (food miles). Slow Food. 19 May. Bra, Cuneo Italy.

Mayring, P., 2000. Qualitative Content Analysis. Forum Qualitative Sozialforschung/Forum: Qualitative Social Research 1. Available at: http://www.qualitative-research.org/fqs-texte/2-00/2-00mayring-e.pdf. Accessed September 2008.

Molteni, L. and Troilo, G., 2003. Ricerche di marketing (Marketing Research). McGraw-Hill, Milano, Italy.

Morgan, D. and Krueger, R., 1993. When to use focus groups and why. In: D. Morgan (ed.), Successful Focus Groups: Advancing the State of the Art, 3-20, Sage, Newbury Park, CA, USA, pp. 3-20.

Panyakul, V., 2003. Thailand's organic status. Ecology and Farming 34: 37-39.

Robson, S., 1990. Group discussions. In: R. Birn, P. Hague and P. Vangelder (eds.), A handbook of market Research Tecniques, Kogan Page, London, UK, pp. 261-76.

Roitner-Schobesberger, B., Darnhofer, I., Somsook, S. and Vogl, C.R., 2008. Consumer perceptions of organic foods in Bangkok, Thailand. Food Policy: 112-121.

Stokes, D. and Bergin, R., 2006. Methodology or 'methodolatry'? An evaluation of focus group and depth interviews. Qualitative market research: An international Journal: 26-37.

Suwannaporn, P. and Linnemann, A., 2008. Consumer Preferences and Buying Criteria in Rice: A Study to Identify Market Strategy for Thailand Jasmine Rice Export. Journal of Food Products Marketing 4: 157-177.

Webb, J.R., 1995. Understanding and Designing Marketing Research. The Dryden Press, London, UK.

Zammuner, V.L., 2003. I focus group. Il Mulino, Bologna, Italy.

Zikmund, W.G., 1997. Exploring Marketing research. 6th ed., The Dryden Press, Fort Worth, TX, USA.

Income effects through trade with organic products for rural households in North Eastern Thailand

U.B. Morawetz[1], R. Wongprawmas[2] and R. Haas[1]
[1]University of Natural Resources and Applied Life Sciences Vienna (BOKU), Department of Economics and Social Sciences, Feistmantelstrasse 4, 1180, Vienna, Austria
[2]Thammasat University, Faculty of Science and Technology, Thammasat University Rangsit Center, 99 Paholyothin Road , Klongluang, Patumtani 12121, Thailand

Abstract

This study groups households in North Eastern Thailand according to their income and degree of specialisation in crop production to derive representative household types. For these household types a linear optimization model is run to calculate net incomes under four scenarios. These are certified organic farming, organic farming in the initial and transitional phase and a self-sufficient farming. Simulations for the different management scenarios show that per ha cash profits are about double under certification while they can only be increased by 30% under self-sufficient farming, even under favourable assumptions. However, transition costs to organic farming are high due to reduced yields at the beginning. According to the figures and model used, only under certified organic production it pays to hire non household workers. Labour hence is a major limiting factor.

1. Introduction

Worldwide there is a growing demand for organic products. In the United States the sales of organic foods are estimated to have grown by 15.7% in 2005 (NFM, 2006) and high growth rates are observed in many other industrialized countries as well. The major consumers of certified organic products are North America, Europe and Japan (Buley *et al.*, 2004). The growing demand is recognized as an opportunity for farmers to increase the value of their products as the consumer's willingness to pay is higher for organic than for conventional products. In many industrialized countries this opportunity to increase the value of agricultural products is supported by subsidies since there is public interest in less environmentally harmful farming. The growing demand and public support has lead to an enormous increase in organic production (Willer and Yossefi, 2006).

Due to the demand for organic products in industrialized countries, organic farming in developing countries has increased as well. Non-Governmental Organisations (NGO) have been engaged for decades to build up producer co-operations and build international trade links. The co-operations often favour organic production since it reduces input costs, is less environmentally harmful, poses less danger to the health of the farmers and can realize higher at the gate prices. These initiatives typically consist of small scale farmers (Oxfam GB, 1994; Udomkit and Winnett, 2002; UN ESCAP, 2002). But with the increasing demand for organic products the production of organic food became interesting for agricultural enterprises as well. A practiced model that allows a maximum of control for the enterprise is contract farming. Farmers become workers on their own land and agree to comply to the agreement with the enterprise (Setboonsarng *et al.*, 2006). As the trade volume increased, traders, certification institutions and governmental bodies that provide support have been established. Additionally, research and education about organic farming is now standard in many agricultural universities around the globe. In the last decade also international organizations and national development agencies increasingly foster organic production in their programs (BMF and ADA, 2006; Buley *et al.*, 2004; UN ESCAP, 2002; Willer and Yossefi, 2005). They see organic farming as an opportunity to reduce poverty while pushing environmentally less harmful farming.

In Thailand, current organic production is mainly rice and some organic vegetables and baby corn. It is estimated that about 0.12% of arable land is used for organic production. A functioning system of governmental and private, IFOAM (International Federation of Organic Agriculture Movements) accredited, certification institutions have been installed in the last decade. For 2006 it is estimated that the total market value of organic products is 20 million US$ which is about 0.12% of total agricultural exports. The majority of organic rice exports goes to the European Union as the organic standards of the United States are not met by many Thai producers (Eischen *et al.,* 2006).

The remainder of this paper is organized as follows. Section 2 provides background about organic agricultural policy in Thailand and about the region this paper focuses on. In Section 3 the objectives of the research are defined. Section 4 gives details about the data and method used to form household types, the mathematical model and its calibration. Finally, the results and concluding remarks are given in the last two sections.

This paper seeks to describe income portfolios of households in North Eastern Thailand and how they can change due to higher at the gate prices through international trade with organic products and through reduction of input costs.

2. Background

In September 2006 a military coup in Thailand over-threw the government of Taksin Shinawatra and the interim-government of Surayud Chulanont was installed by the military. In its economic policy Taksin's government was export oriented and, as critics claim, overspending (The Nation, 2006). As reaction to this and with the experience of the East Asian Crisis in the late 1990s in mind, the interim government of Surayud Chulanont now champions 'sufficiency economy' (originally it was translated as 'self-sufficiency' economy but this was quickly changed after first reactions from the business community (Kanoksilp, 2006). How serious the interim government is about sufficiency economy is manifested through mentioning it in the interim constitution.

The theoretical bases of sufficiency economy was promoted by His Majesty King Bhumipol Adulyadej of Thailand since the 1970's. Since the coupe, sufficiency economy has been widely discussed as the understanding of its practical meaning is unclear to many people. In particular it was emphasized that sufficiency economy is not to be confused with a backward self-sufficiency economy (Noi, 2006). The concept of sufficiency economy is best developed for small scale farms and is known as 'New theory farming'. In a three phase plan the farm first seeks self-sufficiency, in the second phase it forms co-operations with other farms and in the third phase it is involved in trade (Chaipattana Foundation, 2006). Farms following the New theory farming model can manage their farms according to organic farming rules. It is therefore no contradiction to run a certified organic farm that follows the ideas of New theory farming. New theory farming favours, at least in the initial stage, agricultural inputs produced on the farm over inputs bought from the outside. The major difference between New theory farming and organic management is that the former is focused on self-sufficiency while the latter produces for the market.

Self-sufficiency in Thai politics is not a new concept and it has been competing with organic farming for governmental resources already before the coup. In January 2005 the Taksin government approved a national agenda for self-sufficiency. This originally did not include organic farming and it was only included after lobbying from NGOs. But, the proportion of funds devoted to organic farming remained a small share of the total initiative (Eischen *et al.*, 2006). The already weak governmental support for certified organic farming is likely to be even less with the new government's economic focus. It thus remains to business, research and international institutions to promote certified organic farming.

An example for an internationally financed support for organic farming is the research and promotion project for organic agriculture currently done by Thammasat University, Bangkok (Thammasat University, 2005). It is financed by the EU and the aim of the project is to develop organic management methods for rice and cassava in North Eastern Thailand with a minimum usage of external input. It intends to reduce cash costs for fertilizer and increase at the gate prices through certification. The project and this paper focus on North Eastern Thailand where two thirds of Thailand's poor live (Ahmad and Isvilanonda, 2003). Rural poverty in the North East is due to the poor soils, low and unstable precipitation and unstable yields (Entwisle *et al.*, 2005). The main product is rice which is, nowadays, predominantly grown rain-fed with only one harvest per year. In lower elevations (lowland) paddy rice is grown, while on higher elevations (upland) field crops such as cassava, maize or sugar cane are grown (Fukui *et al.*, 2000). The region was sparsely populated until the end of World War II when mortality fell and the population density increased. Under the population pressure rice cultivation expanded from alluvial plans to surrounding terraces. Much of the forest was displaced for upland cash crops, such as cassava. The population pressure was reduced with the increasing use of contraceptives and the construction of roads which allowed migration to the urban centres in the late 1960s (Entwisle *et al.*, 2005). The expansion of agriculture to marginal land made it more vulnerable to weather conditions. But the better connection to the markets that allows off-farm employment and to buy cheap foods helps to buffer these risks (Fukui *et al.*, 2000).

Today income from non-farm sources plays a major role in North Eastern Thailand. A study on household income of three villages in Khon Kaen in North Eastern Thailand used data from 140 households from the years 1995, 1998 and 2002 to analyse the income diversity of households (Ahmad and Isvilanonda, 2003). The figures show that income from rice is, on village average, as low as 15 to 40% of total income. Average income from non-farm and off-farm activities varies between 32 and 63%. The agricultural census of 2003 also suggests an important role of non-agricultural incomes, stating that in North Eastern Thailand 60% of he households live only or mainly from agriculture while 21% live mainly from other sources and another 15% live from agriculture and other sources in equal parts (NSO Thailand, 2003).

3. Data and methodology

Income increases through intensification of rice cultivation like in the Central Province of Thailand is no viable option in the North East with its water scarcity and poor soils. Therefore, organic agriculture is seen as a way to increase the value of the production in this area and the calculations in this study all refer to this region.

3.1 Household types

For household level data a large scale survey from University of Chicago (Townsend, 1997) is re-used. The so called 'Townsend project' collected data from North Eastern Province in 1997 just a couple of months before the Asian Economic crisis began with the devaluation of the Thai Bath. In the North Eastern Province two regions (Buriram and Srisaket) were chosen for data collection as for those two, benchmark data were available. Within each of these two provinces

12 tambons (administrative units) were selected by using stratifications by land cover classes from satellite imagery (Binford *et al.*, 2004). Within each tambon, four villages were selected at random. From the Community Development Department's enumeration list 15 households in each village were randomly selected. In total about 1,400 households were interviewed in North Eastern Thailand in May 1997.

To derive representative households we followed the example of Ellis (2000) and grouped them according to their income and sources of income: the households sampled are divided into three groups, depending on their income. The lowest net income is below 17,000 baht per year, the middle group lower than 43,896 baht and the highest above this value (Exchange rate: baht/euro: 46.6, 2007). Figure 1 shows the distribution of the income where the fat vertical lines indicate the thresholds between the lower, middle and upper third of the income. There is a long tail with high income households while the vast majority earns a far lower income. The income is per household and the study is therefore limited to income flows and does not deal with the much more complex issue of poverty. As can be seen from the figure, income can also be negative in a particular year.

For each of these three groups net income portfolios are calculated to quantify the sources of income. From Figure 2 the shares of average net income from different sources for households grouped by income can be seen. Income from 'crop' is primarily from rice, 'off-farm' work consists of all income through paid work (also agricultural work on other farms), 'remittances' contain transfers from (migrated) relatives, government transfers etc. while the category 'other' contains incomes from renting out tools and those incomes specified as 'other' in the survey. Incomes through changes in the assets (e.g. selling of land) were excluded. The figure shows that the main sources of income are from crop cultivation, off-farm work and remittances. 'Livestock' and 'other' have only a minor contribution to income. The differences between household groups are not very big, though, for the high income group off-farm work plays a more important role.

The grade of specialization on crop cultivation is used as criteria to form household types within each of the three income groups. The four types have (1) no, (2) more than zero but less than

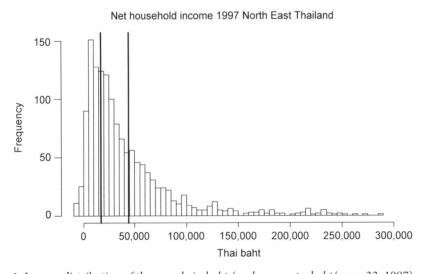

Figure 1. Income distribution of the sample in baht (exchange rate: baht/euro: 32, 1997).

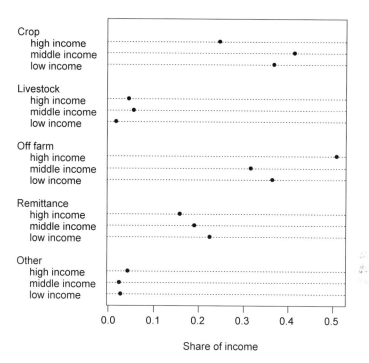

Figure 2. Average shares of sources of income for household groups.

one third, (3) more than a third but less than two thirds and (4) more than two thirds of their net income from crops. In total this results in 12 types. Table 1, describes some characteristics of these household types.

In order to mitigate the impact of outliners, the medians are used to describe the groups. Of the lowest income third, almost half (14.7% of the total population) has no income from crops.

This type is also worst off in terms of most other assets: the median of the maximum of years a family member spent at school is far lower than that of other households. They own less land (but it is positive as this can also be the plot where their house is built) and they have less agricultural and household assets than every other household. Also their social network in terms of relatives and the persons older than 18 years in the household are slightly lower than other households. Only in terms of debts per income they are in a better situation than the farming households'. What is true for the lowest income third, is not true, for the middle and high income third: there the households without income from crops are not worse off than their farming counterparts. This suggests that there is a better educated group of households specializing on non-farm jobs. Worth to mention is that no household type has savings, some of them even substantial amounts of debts.

Comparing the households with the same share of income from crops, it can be seen that those with higher incomes have more or equal education, more land they own, more relatives, more household and agricultural assets and more family members over 18 years of age. It thus can be assumed that factors as education, agricultural assets, land ownership and the social network contribute to the income.

Table 1. Population share, median income (1997 baht per year)[1] and median characteristics of household types.

Income	Crop %	Population %	Income (baht)	Sav./inc.	Max. school[2]	Land owned (ha)	Household assets	Agricultural assets	Relatives
Low	0	14.7	8,200	-0.05	4	0.12	65,000	0	12
Low	<33	5.4	12,480	-0.15	6	1.44	86,600	10,000	15
Low	<66	4.9	11,010	-0.22	6	1.56	105,550	10,500	14
Low	>66	7.4	9,849	-0.14	6	2	91,400	20,000	13
Middle	0	7.3	28,530	-0.01	6	0.16	102,050	0	12
Middle	<33	7.4	27,170	-0.11	8	1.68	125,000	21,500	15
Middle	<66	10.1	27,400	-0.14	7	2.56	142,900	27,000	15
Middle	>66	8.9	27,390	-0.17	7	2.9	147,900	48,500	14
High	0	8.5	72,000	-0.03	8	0.16	164,800	0	13
High	<33	12.5	82,160	-0.03	9	3.46	216,850	31,500	15
High	<66	7.1	66,710	-0.13	8	4.64	204,500	52,000	16
High	>66	5.8	60,700	0.00	9	4.8	206,400	57,000	15

[1] Exchange rate: baht/euro: 32, 1997.
[2] Max. school: years spent in school.

3.2 Mathematical model

A linear one period farm level optimization model is used to describe the household's behaviour under different scenarios (a mathematical summary is given below). Households maximize cash income by choosing the management method and working off-farm. Income consists of three sources: (1) crop cultivation, (2) off-farm work and, (3) the remaining income sources (consisting of remittances, livestock and other and noted with an $R^{observed}$). The latter is not modelled but just assumed to be fixed in the short run. The income from off-farm work is subject to an upper limit which is set to be the observed value $NF^{observed}$. The idea behind it is that labour markets allow only a certain level of employment in off-farm activities in rural areas. It is assumed that households already work off-farm as much as the labour market allows.

The income from crops is generated through cultivation (x_c gives the ha planted with crop c) and sales (s_c gives the sales in kg) minus labour (V_t^T gives the hired workers days and r_t^T the wage) and input costs (m_c are the input costs for crop c per ha). The model also allows production of cassava on upland fields. But, as shown below, the median values revealed no upland fields for the median households (this is surprising as upland crops are considered to be an important cash source for households). Farmers can opt between organic and conventional management of their crops (modelled as different crops c). They can employ workers and choose how much off farm work they do (h_t^{off} are the days in month t and w_t^T is the wage for off farm work). Since here only cash flows are modelled, family farm work is supplied at a wage of zero.

As a property of linear models, the optimization algorithms do not choose mixed strategies. The model therefore opts for the choice with the highest marginal income. The wage for off-farm work is set marginally lower than the costs for hired workers. Thus family members prefer to do the farm work themselves. Mathematically, the model can be summarized in the following equation (Equation 1):

$$\max \left(\sum_c p_c s_c - r^T \sum_t V_t^T - \sum_c m_c x_c + \sum_t w_t d_t + R^{observed} \right) \tag{1}$$

subject to

$$\sum_c l_{tc} x_c \leq L \qquad \forall t$$

$$\sum_c v_{tc} x_c \leq h_t^{farm} \, V^F + V_t^T \qquad \forall t$$

$$y_c x_c = s_c$$

$$h_t^{farm} + h_t^{off} \leq V^F d$$

$$\sum_t w_t d_t \leq NF^{observed}$$

$$x_c \geq 0 \qquad \forall c$$

$$V_t^T \geq 0 \qquad \forall t$$

$$s_c \geq 0$$

where y_c is the yield per ha, p_c is the price of crop c, L is the total land available in ha, l_{tc} is the fraction of a month that crop c occupies the land, v_{tc} is the labour required for crop c during moth t, and d are the working days per month per person.

3.3 Calibration

The data for the endowment of the different household types and the input costs are derived directly from the Townsend project dataset. Other necessary data had to be taken from other research or had to be assumed. Monetary values are calculated in 2003 Thai baht.

The data of median endowment and median input costs derived from the Townsend project data are provided in Table 2 and Table 3. Monetary values are multiplied by a factor of 1.24 to adjust for inflation in the years between 1997 and 2003 (BTEI Thailand, 2004). It can be seen that households with higher income also cultivate more lowland if compared with households with the same grade of specialisation. The median of the persons older than 18 years is two for the households with low income while it is three for the households with middle or high income. The medians for the input costs in Table 3 are not as easily structured. Dummy variable regressions also did not show a significant influence of the 'type of household dummy' on the costs. But, in another regression, the size of cultivated lowland could be shown to have a significant influence on the costs per ha (regression not shown here). In absence of better data, the medians presented in Table 3 were used as cash input costs per ha.

The Townsend project data provide no data on yield. Therefore results from a working paper of the Asian Development Bank which builds on a data collection from Ubon Ratchathani, Surin and Yasothon in North Eastern Thailand in 2003 are used (Setboonsarng *et al.*, 2006). According to this study, rice yields per ha are 2,181 kg/ha on average, and the price per kg of conventional

Table 2. Persons and land endowments used in the model and sources of income in%.

Income	Crops %	Persons >18	Upland ha	Lowland ha	Off-farm income	Crop income	Other income	α
Low	<33	2	0	1.12	0.74	0.26	0.00	0.22
Low	<66	2	0	1.02	0.28	0.72	0.00	0.64
Low	>66	2	0	1.92	0.00	1.00	0.00	1.07
Middle	<33	3	0	1.28	0.67	0.28	0.05	0.52
Middle	<66	3	0	1.92	0.34	0.61	0.05	1.06
Middle	>66	3	0	2.3	0.00	1.00	0.00	2.01
High	<33	3	0	2.08	0.81	0.19	0.00	0.99
High	<66	3	0	3.2	0.28	0.63	0.09	2.67
High	>66	3	0	3.36	0.02	0.98	0.00	4.1

Table 3. Cash input costs per ha in baht (2003)[1].

Income	Crops %	Fertilizer	Pesticides	Seeds	Machines	Total
Low	<33	1,875	0	0	750	2,625
Low	<66	1,817	0	0	516	2,332
Low	>66	1,250	0	0	573	1,823
Middle	<33	1,479	0	0	670	2,150
Middle	<66	1,432	0	0	625	2,057
Middle	>66	1,606	70	0	747	2,422
High	<33	1,559	10	0	694	2,264
High	<66	1,330	0	0	497	1,827
High	>66	1,382	20	0	500	1,902

[1] Exchange rate: baht/euro: 45, 2003.

rice is 5.87 baht/kg. For farm workers the study suggests a wage of 195 baht/person/day for contract farms.

Working hours per ha per month are taken from Fukui (1993: 223). In his studies he observed families during their peak working times. This is used as guidance for the work effort during different months. The last 12 columns of Table 4 give the work effort for different months for rice cultivation. Finally, for family members a working month is assumed to have 25 working days.

For calibration of the crop activities, first a base scenario is run. The calculated income from crop cultivation is compared with the observed income and the factor by which they differ is used to scale the calculations in the following scenarios. This factor is called alpha and corrects the model for errors due to misspecification. An alpha smaller than 1 means that the observed income from crops is lower than the results of the model and an alpha greater than 1 means that the observed income from crops is higher than the model outputs. An implicit assumption is that alpha does not change if another management is applied. The values for alpha are given in the last column of Table 2. The correlation between the ha cultivated and alpha is 0.93. This indicates

Table 4. Key data of different management scenarios.

	Yield kg/ha	Price %	Cash cost %	Work days per month											
				1	2	3	4	5	6	7	8	9	10	11	12
Base	2,181	100	100	0	0	0	0	0	10	35	35	5	3	50	50
Organic	2,181	170	60	10	10	10	11	20	20	35	35	18	5	50	50
Transition	1,636	122	60	10	10	10	11	20	20	35	35	18	5	50	50
Initial	1,091	107	60	10	10	10	11	20	20	35	35	18	5	50	50
Sufficient	2,181	100	10	10	10	10	11	20	20	35	35	18	5	50	50

that without alpha the income from crops for small scale farms is overestimated and for larger farms underestimated.

3.4 Management scenarios

Different management scenarios are applied to the model described above. The scenario 'base' describes the income under conventional management. It is used to derive alpha. The scenario 'organic' describes a household that cultivates organic rice along the guidelines of an organic certification organization and receives a substantially higher yield at the gate price. Also input costs are reduced. Yields are as high as under conventional management as farmers are experienced in organic cultivation. The scenario 'transition' describes a farm that has been under organic management for two to four years but has not yet been certified. The at the gate price is not as high as for 'certified' farms but input costs are reduced. Yields are lower since the soil has not yet fully recovered from chemical fertilisation and the farmer is not as experienced. The scenario 'initial' is for farms that have their first or second year of organic management. The at the gate price is only slightly higher than for conventional products but yields are reduced even more than for farms in transition. The last scenario is not about organic management but is a stylized version of 'sufficiency' economy. It is only the first step of the three steps suggested by New theory farming in which the farm seeks to reduce dependency by reducing input costs. The way it is modelled here, the work effort is as high as for organic farming but at the gate prices are not as high as for organic products. Yields are as high as under conventional management.

The data for the different scenarios are collected from various sources. The prices for organic farming are taken from the above mentioned Setboonsarng *et al.* (2006) paper. Table 4 shows that even in the initial phase and during transition farmers have slightly higher prices compared to conventional farming. This is possible by selling them on local markets as 'pesticide save'. Several studies showed that yields under organic management can be as high as under conventional farming (Khunthasuvon *et al.*, 1998; Setboonsarng *et al.*, 2006). The Setboonsang *et al.* study even suggests that during the initial and the transitional phase yields are not significantly different. But since samples size is very low and other evidence suggests that yields are temporarily reduced, we assume that yields are reduced by 50% during the initial phase and by 25% during the transition phase (a less arbitrary determination of yields would require assumptions about many agricultural parameters which are not available in this model). Cash costs for conventional farming depend on the household type. For organic farming scenarios it is assumed that cash cost can be reduced due to abandonment of pesticides (100% reduction), reduced fertilizer cash costs (47% reduction) and lower cost for machinery (24% reduction). These reductions are derived from the reductions in cash costs given in Setboonsang *et al.*. Depending on the use of pesticides in the base scenario this is a total reduction of input cost between 40 to 42% in comparison to the base scenario. The working hours for organic management are higher as is shown in Table 4. Here, once again, it is difficult to find appropriate values in scientific publications. The values given are pure assumptions. Note, that work effort is identical during peak working seasons in all scenarios. If it was assumed that work effort during peak seasons was higher under organic or sufficient management, this would reduce profits as labour had to be hired.

The figures for the sufficiency scenario are a combination of the price for conventional rice and work effort for organic management. But it is assumed that input costs are reduced by 90%. It is unclear if such a high reduction is possible while keeping the yield on the level of conventional farming, but for the sake of argument exaggeration it is preferred in this context.

4. Results

Column 'Base' of Table 5 shows cash profits from one ha of rice cultivation. It is between 9,549 baht per ha for the households with low income (with crops <33) to 10,538 baht per ha for households with high income (with 34<crops <66).

Roughly, households that plant more crops, have higher cash profits per ha (correlation of 0.64). The following 4 columns of Table 5 give the percentage change in cash profit per ha for the four scenarios. With an increase of the price by 70% and input costs reduced by about 40%, cash profits per ha for certified organic management are increased by about 100% for all household types. In the transition phase, when cash input costs are reduced by about 40%, but yield is reduced by 25%, cash profits per ha remain approximately the same as in the base scenario for all household types. In the initial phase a reduction of cash profits per ha of 48% is calculated for all households. This is due to the yield reduction by 50%. For the sufficiency scenario, income per ha is increased between 19 and 31%. In particular households with high input costs, which are predominantly those not specialized on crop cultivation, gain in the sufficiency scenario as they can reduce their expenses in inputs most.

Table 5. Base profit in baht (2003)[1] per ha and changes in percentages under different management.

Income	Crops %	Base	Organic	Transition phase	Initial phase	Sufficient
Low	<33	9,549	108	2	-48	31
Low	<66	9,912	103	1	-48	26
Low	>66	10,544	94	-2	-48	19
Middle	<33	10,138	99	-1	-48	24
Middle	<66	10,253	98	-1	-48	22
Middle	>66	9,800	105	1	-48	28
High	<33	9,996	101	0	-48	25
High	<66	10,538	94	-2	-48	19
High	>66	10,446	96	-1	-48	20

[1] Exchange rate: baht/euro: 45, 2003.

A transition to organic agriculture is, according to these calculations, costly in the initial phase and valuable after certification. Even though a reduction of input cost by 90% was assumed, in the sufficiency scenario per ha profit increases only up to 31%.

Table 6 gives the total household profits in the base scenario and the changes in percentages under different management strategies. In the different scenarios households are forced to apply the respective management, even if it does not maximize profit. The results are similar to those from Table 5. Trivially, households without income from crops are not affected from the management decision. But, as mentioned already above, this is the type with the lowest income. The main winners from certified organic crop production are households whose main income is from crop production. But, if they want to change to organic management all at once, they also face the highest absolute losses during the period of transition.

Under the conventional, initial, transitional and sufficient management scenarios, only family labour is used for cultivation since marginal profits are not high enough to employ workers. This limits the cultivated area to the working capacities of the household during the peak working season. With organic farming, having a higher marginal profit, it becomes feasible to hire workers. The increased cultivated area increases households profits under organic farming more than proportional. Never the less, even under organic management labour is a limiting factor as cost per ha keep on raising with the area cultivated. Figure 3 shows the average cash profits per ha for high and low income households with more than 66% of their income from agriculture under organic and conventional management. The differences in profits between organic and conventional farmers are substantially and are mainly due to the higher price of organic rice. But the striking fact about cash income is the quick reduction as soon as workers have to be employed. Richer households have more household members and they can therefore cultivate more at lower cash costs.

Table 6. Household income in baht (2003)[1] and changes in percentages under different management.

Income	Crops %	Base	Organic	Transition phase	Initial phase	Sufficient
Low	0	4,400	0	0	0	0
Low	<33	8,327	0.28	0.01	-0.12	0.08
Low	<66	8,866	0.74	0.01	-0.34	0.19
Low	>66	11,284	0.98	-0.02	-0.48	0.19
Middle	0	22,000	0	0	0	0
Middle	<33	23,808	0.28	0.00	-0.14	0.07
Middle	<66	26,784	0.61	-0.01	-0.29	0.14
Middle	>66	29,601	1.07	0.01	-0.48	0.28
High	0	62,400	0	0	0	0
High	<33	79,174	0.20	0.00	-0.09	0.05
High	<66	67,101	0.62	-0.01	-0.30	0.12
High	>66	65,993	0.97	-0.01	-0.46	0.20

[1] Exchange rate: baht/euro: 45, 2003.

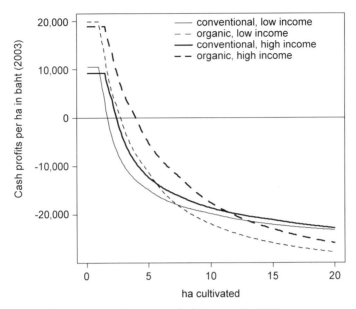

Figure 3. Average profits per ha. Exchange rate: baht/euro: 45, 2003.

5. Final remarks

The results presented are derived from a calculation of cash incomes of different household types. The results allow several conclusions which can contribute to the discussion about agricultural policy in Thailand.

The data used show that households with the lowest income have no income from crops. Households with higher income gain in absolute terms more from organic agriculture as they cultivate more land. Organic farming is therefore no policy that favours households with low incomes (directly). Nevertheless, in relative terms, households with low income but a high specialisation in agriculture can gain substantially. Differences in per ha cash input costs of different household types do not have a magnitude that plays a decisive role in which management system to choose.

According to the data, a reduction of cash input costs by 90% can increase cash profits by up to 31% while, under organic farming, cash profits per ha increases of 100% are possible. International trade with its high price premiums therefore allows increases of income by much more than what can be achieved through reduction of input costs.

Available labour is a limiting factor as wages are relatively high. According to the figures used in this model, cultivation does not pay if labour has to be hired. Under organic management, per ha profits are high enough to hire workers to increase the cultivated area. Labour scarcity is limiting the cultivated area in particular during planting and harvesting time. Organic farming techniques are therefore more suitable if they do not increase the work effort during the peak working season. This raises the question to which degree labour can be substituted by capital in organic agriculture and if a more intensive organic agriculture could be a way out.

Yield reductions during the initial phase of organic farming make it expensive to change to organic management. In particular households that have the majority of their income from crop cultivations can suffer high losses which might not be affordable as many households have debts already. It is therefore critical to keep yields high during the years before certification which is possibly achievable through training and research.

Discussion in lecture room

Apply your theoretical knowledge with respect to this paper:
- This paper applies a very simple approach to calibrate the model. Explain in your own words why calibration is necessary.
- To describe household characteristics in Table 1 the authors used the median to describe the distribution of the variables of interest. Why do you think they preferred this measure instead of the arithmetic mean?
- In Figure 3 you see kinks in the curves describing cash profits. Can you tell how these kinks come about? Asked differently, which restriction turns into a binding restriction at the kinks?

Acknowledgments

This paper is based on research performed in the framework of the BEAN-QUORUM project, co-funded by the European Commission through the Asia-Link Programme. This research was possible through a scholarship financed by the EU Asia Link Program BEAN-QUORUM. Special thanks for their help to my friends at Thammasat University in Thailand.

References

Ahmad, A. and Isvilanonda, S., 2003. Rural Poverty and Agricultural Diversification in Thailand. Second Annual Swedish School of Advanced Asia and Pacific Studies (SSAAP), 24-26 October, Lund, Sweden.

Binford, M.W., Lee, T.J. and Townsend, R.M., 2004. Sampling design for an integrated socioeconomic and ecological survey by using satellite remote sensing and ordination. Proceedings of the National Academy of Sciences of the USA 101(31): 11517-11522.

BMF and ADA, 2006. Guidelines for targeted funding of CGIAR Centers. Vienna, Austria, Federal Ministry of Finance of Austria and Austrian Development Agency: 21.

BTEI Thailand, 2004. Consumer Price Index 1994 - 2004, Bureau of Trade and Economic Indices; Ministry of Commerce Thailand.

Buley, M., Jährmann, K., Kotschi, J., Richert, A. and Spitta, J., 2004. Ökologische Landwirtschaft. Ein Beitrag zu einer fairen Globalisierung. Frankfurt am Main, Germany, Gesellschaft für Technische Zusammenarbeit: 40.

Chaipattana Foundation, 2006. The New Theory, Chaipattana Foundation, Thailand.

Eischen, E., Prasertsri, P. and Sirikeratikul, S., 2006. Thailand. Organic Products. Thailand's Organic Outlook. Bangkok, Thailand, Global Agricultural Information Network: 25.

Ellis, F., 2000. Rural Livelihoods and Diversity in Developing Countries, Oxford University Press, UK.

Entwisle, B., Walsh, S.J., Rindfuss, R.R. and Vanwey, L.K., 2005. Population and Upland Crop Production in Nang Rong, Thailand. Population and Environment 26: 449-470.

Fukui, H., 1993. Food and Population in a Northeast Thai Village., University of Hawaii Press, Honolulu, Hawaii.

Fukui, H., Chumphon, N. and Hoshikawa, K., 2000. Evolution of rain-fed rice cultivation in northeast Thailand: increased production with decreased stability. Global Environmental Research 3: 145-154.

Kanoksilp, J. 2006. Sufficiency economy difficult to swallow. The Nation, Bangkok, Thailand.

Khunthasuvon, S., Rajastasereekul, S., Hanviriyapant, P., Romyen, P., Fukai, S., Basnayake, J. and Skulkhu, E., 1998. Lowland rice improvement in northern and northern Thailand 1. Effects of fertiliser application and irrigation. Field Crops Research 59: 99-108.

NFM, 2006. NFM Market Overview. Natural Food Merchandiser 27: 16-18.

Noi, C., 2006. What does Sufficiency Economy mean? The Nation, Bangkok, Thailand.

NSO Thailand, 2003. Agricultural Cenus 2003 Northeastern Region. National Statistical Office and Ministry of Information and Communication Technology Thailand.

Oxfam GB, 1994. Program Impact Report. Oxfam GB's work with partner and allies around the world., Oxfam GB: 74.

Setboonsarng, S., Lueng, P. and Cai, J., 2006. Contract Farming and Poverty Reduction: the Case of Organic Rice Contract Farming in Thailand, Asian Development Bank Institute Discussion Paper No. 49.

Thammasat University, 2005. Enhancing competitiveness of organic rice and tapioca cultivations to stimulate local development and Thai-EU trade relations. EU-Thailand Economic Co-operation Small Project Facility Grant Application Form. Thammasat University, Faculty of Science and Technology, Bangkok, Thailand.

The Nation, 2006. New Economic Direction, Call for definition of self-sufficiency. The Nation, Bangkok, Thailand.

Townsend, R.M., 1997. Townsend Thai Project Initial Household Survey. Social Science Computing Service, the University of Chicago, USA.

Udomkit, N. and Winnett, A., 2002. Fair Trade in organic rice: a case study from Thailand. Small Enterprise Development 13: 45-53.

UN ESCAP, 2002. Organic Agriculture and Rural Poverty Alleviation Potential and best practice in Asia. UN ESCAP, Bangkok, Thailand.

Willer, H. and Yossefi, M., 2005. The World of Organic Agriculture. Statistics and Emerging Trends 2005. IFOAM, Bonn, Germany.

Willer, H. and Yossefi, M., 2006. The World of Organic Agriculture. Statistics and Emerging Trends 2006. Bonn, IFOAM, FiBL, Germany and Frick, Switzerland.

An overview of the organic and green food market in China

S. Marchesini[1], H. Hasimu[1,2] and R. Spadoni[1]
[1]Alma Mater Studiorum University of Bologna (DEIAgra), Department of Agricultural Economics and Engineering, Viale Fanin 50, 40127 Bologna, Italy
[2]Xinjiang Agricultural University, No.42 Nanchang Road, Urumqi, Xinjiang, China

Abstract

This paper presents an overview of Chinese organic and green food markets, examining historical development, production bases, market condition and certification systems, in order to point out issues that are most likely to affect future development, from the point of view of international standardisation. What emerged is that domestic consumption of organic food is growing in China, partly attributable to worries over food safety, but exports (mostly raw materials) are the major reason for growth, and yet organic remains a tiny niche market. The factors that slow down the growth process in the domestic market include low public awareness (both among consumers and producers), lack of specialized technical knowledge about organic production, the poor reputation of law enforcement, as well as the harmonization of standards and certification procedures, not yet fully compatible.

1. Introduction

Arable land is a precious resource for China's agriculture. Most of the country is covered by mountains, deserts, or dry grasslands, all unsuitable to agriculture. By the end of 2006, China had about 122 million hectares of arable land, covering 13% of its territory; this amounted to 0.09 hectares per capita, less than 40% of the world per capita average, 1/8 the U.S. level, and one half the Indian level. Furthermore, China's population has been growing by some 10 million people annually, and arable land is being lost to new construction, natural disasters and conversion of farmland to other purposes (e.g. lower-quality arable lands are used for forest or grassland replanting), not to mention the pollution and soil erosion problems that plague the remaining farmlands; an efficient agriculture policy is therefore a top priority for the country, for it directly affects national food security. Starting in the 1980s, the Chinese government pushed the adoption of technologies that maintain high food production, so GMOs (genetically modified organisms), fertilizers and pesticides have been widely used as a means to increase yields. China is one of the world's biggest producers, users, and exporters of agro-chemicals, and although most highly toxic chemical substances were banned from agricultural use in early 2007, farmland contamination is still a relevant issue (Baer, 2007; Yang, 2007). The production boost came along with a major cost to the environment and consequently to human health. The turn towards organic was initially fostered by the emergence of safety issues regarding Chinese agricultural exports. For instance, the European Union (EU) and Japan banned tea and other commodities imported from China due to a level of pesticide residues exceeding the threshold allowed by their food safety standards. Moreover, the attention of Chinese consumers' towards food quality has increased remarkably over time, in part due to the impact of misuse of chemical inputs on consumers' health – several thousand people are poisoned by pesticides every year (OCA, 2003) – in part due to the recent food crises and scandals (e.g. milk, fish, pork, eggs, and chicken contaminated with the industrial chemical melamine), which highlighted China's woeful food safety standards. Urban consumers' income growth (Veeck and Veeck, 2000), combined with food security concerns, raised the demand for quality food, safer production, and processed foods, and attracted more and more the attention of the media to this issue. The XI five-year plan (2006-2010) itself aims at constituting a 'harmonious society' relying on advanced science and technology to achieve high-quality and high-efficiency development, and seeks sustainability, advocating higher living standards for human beings as a reaction to past misuse of resources. Organic agriculture is environmental friendly, sustainable and health-oriented, and although its natural methods conflict with concepts of food production maximization, it is now strongly supported by Chinese government. The creation of 'organic zones' in the area around Beijing, while being aimed at taking advantage of the growing demand for organic food in that area, is an attempt by the central political powers to propose a model for other cities' governments (Baer, 2007).

2. China's ecological food quality labels

Quality labels are instruments which allow the identifying of a product according to some specific features, such as their origin (e.g. PDO/PGI), production methods (e.g. organic), or residue contents (no residues). Labels play a relevant role in both domestic and international markets, as they guarantee that the declared characteristics are effectively possessed by the product. Quality labels are distinctive signs of the nature of the products, and along with quality certification they encourage the qualitative transparency of the foods. Obviously, the use of a label implies more costs, thus causing the selling price to rise in turn.

Three types of food ecological quality certification and labelling systems are currently available in China (Zhang, 2005):
1. pollution-free (safe, hazard-free) food;
2. green food;
3. organic food.

China's total agricultural land is about 122 million hectares; the eco-labelled food production area in 2007 is 34.18 million hectares, or 28% of the total (Paull, 2008a). Figure 1 shows the breakdown of the agricultural food production area.

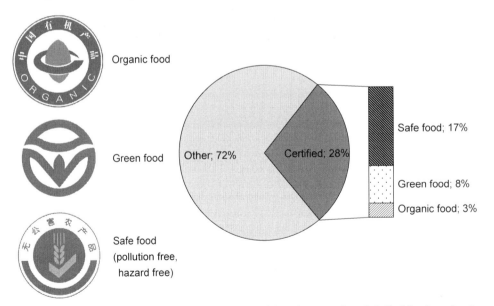

Figure 1. Chinese food quality certification signs and distribution of eco-labelled food production as a proportion of the total cultivated land in China in 2007 (SCIO, 2007).

2.1 Safe food

The Safe food program was launched in the 1980s and became operative in 2001. Safe food (translated also as 'hazard free' or 'pollution free') is guaranteed under a system focusing on the environment of farming land, production process and quality. It has a wider distribution because of its low price, and it is easier for farmers to produce because the standards are less strict. Certified, safe food is not regarded by consumers in the same way as organic or green food, although most consumers do not understand the distinction between organic and green food rating-systems.

2.2 Green food

The Green food program was initiated in 1990 by the Chinese Ministry of Agriculture, and in the following year its label was registered as the first certification of this type in China. The China Green Food Development Centre (CNGFDC) owns the registered label and also acts as certification body. The products that may be labelled as green food cover 7 major categories and 29 sub-categories of agricultural and food products, including grain, oil, fruit, vegetables, animals (meat, dairy, and aquatic) products, alcoholic and non-alcoholic beverages. The popularity of green food labelling has soared and by the end of 2007 green food production was valued at 20.7 billion US$ (Paull, 2008a: Figure 5). A survey in Beijing supermarkets revealed that consumers show a positive attitude and are familiar with green food (Xia and Zeng, 2007).

As noted by Paull (2008b), Green food provides a 'middle way' between conventional and organic farming. Green food is graded AA and A. AA-grade green food requires, unlike standard A-grade green food, traceability and the absence of any synthetic agro-chemicals residue. It is comparable to, but differs from, organic products since it depends on product standards rather than process standards as organic products do. In this respect, the green food quality assurance and certification process makes extensive use of modern test methodologies to ensure that the production environment and the characteristics of the final products meet its benchmarks. Green foods have an end-product and food safety orientation, aimed at meeting the needs of consumers and government.

The green food model puts emphasis on initial field tests and on laboratory tests of products, while the field inspection is not as traceable as that of organic food. The inspection of standard A-grade green food relies more on the production and control records of green food enterprises, while the inspection of AA-grade green food products is more process-oriented and therefore more similar to that of organic agriculture.

The Chinese government promotes green food because of it is believed to play a role in improving overall food quality. However, it is a solely Chinese certification, therefore when exported it is marketed as conventional. The demand for green food in Japan, for instance, is primarily driven by the expectation that it is also compliant with the basic import requirements of such markets, while China's regular food exports may not be able to meet several developed countries' standards (IFAD, 2005).

According to Lu (2002), the milestones in the development of the organic food sector in China can be summed up as follows, described in 3 stages:
- Stage 1 (1990-1993): start-up stage. The main responsibility for implementing the Green food project is assigned to the China Green Food Development Centre (CNGFDC). Quality testing

and controlling agencies were founded, and quality criteria and standards were defined. Green food production grow quickly but in selected areas only.

- Stage 2 (1994-1996): rapid development stage. This stage is characterized by a significant increase in numbers of products, acreage and production.
- Stage 3 (1997-2007): maturity stage. Green food becomes popular among the public, and develops through marketing and internationalization. The 'socialization' of green food became a reality through the following developments:
  - Consumers and media acknowledge green food, and become familiar with it.
  - Green food is marketed in many large and medium size cities.
  - Local government devote more attention to green food development.
  - Specific technologies are developed for green food production.
  - Large enterprises are involved in green food.
  - Green food market expands. From 1997 to 2007 the number of certified products increased from 892 to 14,229; enterprises grew from 544 to 5,315; output grew from 6.3 to 72 million tons; hectares expanded from 2.14 to 10 million hectares; production value grew from 2.9 to 20.7 billion US$; export value grew from 70 million to approximately 2 billion US$.

2.3 Organic food

The term 'organic' is best thought of as referring not to the type of inputs used, but to the concept of the farm as an organism, in which all the components – the soil minerals, organic matter, micro organisms, insects, plants, animals and humans – interact to create a coherent, self-regulating and stable whole. Reliance on external inputs, whether chemical or organic, is reduced as much as possible (Lampkin *et al.*, 1999).

For this reason, the organic food approach (process oriented) is rather different from that of green food (more product oriented). The certification process for organic farmers and food producers do not regularly require environmental or sample tests unless problems are suspected, while the certification process implies traceability and follows the whole production process for each organic food back down to the individual farmer and single plot of land. In addition, rather than focusing just on safety needs, organic farming historically developed from a mixture of philosophical, environmental, and safety needs and is more oriented to comply with farmers' expectations. Organic farmers are required to employ active measures to seek to improve the environment or to reduce the impact of their activities. If sold on the market at a premium, organic production thus internalizes public benefits (such as lower pollution, biodiversity preservation and other environmental aspects). These benefits are bundled in a product and by buying this product consumers directly provide support to this value-enhancement strategy, thus creating a market incentive for farmers to produce such environmental services (IFAD, 2005).

Worldwide, more than 32 million hectares have been certified for organic agriculture in 2007 (Willer and Kilcher, 2009); if wild harvest areas are also included then the certified organic area reaches 51.1 million hectares according to the International Federation of Organic Agriculture Movements (IFOAM). In total, Oceania holds most of the world's organic land, followed by Europe and Latin America (Table 1). The proportion of organically- compared to conventionally-managed land, however, was highest in Europe, while Latin America had the greatest total number of organic farms (Willer and Kilcher, 2009).

In 2007, the countries with the greatest organic areas were Australia (12 million hectares), Argentina (2.8 million hectares), Brazil (1.8 million hectares), USA (1.6 million hectares) and China (1.6 million hectares) (Table 2). Most of the organic agricultural land in China is

Table 1. Organic agricultural land and producers by region 2007 (including in conversion areas) (Willer and Kilcher, 2009: shares are based on the total agricultural land of the countries included in the survey).

	Organically managed agricultural land[1]		Share of total agricultural land[1]	Producers	
	Hectares	%	%	Number	%
Africa	870,329	2.7	0.1	529,986	43.5
Asia	2,881,745	8.9	0.2	234,147	19.2
Europe	7,758,526	24.1	1.9	213,297	17.5
Latin America	6,402,875	19.9	1.0	222,599	18.3
North America	2,197,077	6.8	0.6	12,275	1.0
Oceania	12,110,758	37.6	2.6	7,222	0.6
Total	32,221,310	100.0	0.8	1,219,526	100.0

[1] Excluding aquaculture and wild collection areas. Includes in-conversion areas.

Table 2. Area under organic management by country in 2007 (Willer and Kilcher, 2009: shares are based on the total agricultural land of the countries included in the survey).

Rank	Country	Hectares	% organic	Number of producers
1	Australia	12,023,135	2.70%	1,438
2	Argentina	2,777,959	2.15%	1,578
3	Brazil	1,765,793	0.67%	7,250
4	USA	1,640,804	0.51%	8,493
5	China	1,553,000	0.28%	1,600
6	Italy	1,150,253	9.05%	45,231
7	India	1,030,311	0.57%	195,741
8	Spain	988,323	3.93%	18,226
9	Uruguay	930,965	6.23%	630
10	Germany	865,336	5.11%	18,703

permanent pasture; permanent crops do not play a major role. Arable land is mainly used for cereals, including rice. As far as the organic area as a percentage of the total agricultural area is concerned, China ranked 69th; counting organic farms China ranked 55th.

The amount of agricultural land under organic management increased enormously in China from 2000 to 2006, and in 2006 it was accredited by the IFOAM survey for 3,466,570 hectares. In the year 2005/2006, China added 12% to the world's organic area. This accounted for 63% of the world's annual increase in organic land, and allowed China to reach 11% of the world's organically-managed land. According to Paull (2007), China has adopted an innovative path, via Green food, towards achieving an organic future. However, in 2007 the reported area under organic management was smaller (1.55 million hectares according to Willer and Kilcher, 2009). This is due to the fact that the Certification and Accreditation Adminstration of China (CNCA) excluded farms certified by foreign certifiers for compliance with foreign standards and also excluded bee pastures, extensive pastures and other wild collection areas.

China is a large country, and rich in natural resources, which gives organic food producers a large range of choices among crops and food products. Many places in China are also rich in faunal resources. These biological resources always have special local flavours and are good for health, and therefore will have wide market prospects.

On the other hand, however, China is considered a highly polluted environment (WHO, 2007). According to the World Health Organization report on air quality, 7 out of 10 of the most polluted cities in the world are in China. The Chinese government itself recognises that two thirds of Chinese cities suffer from air pollution, and that the vast majority of the population drink contaminated water. In addition, in recent years, the issue of China's food safety has gained a prominent position both in the international and domestic media due to the recurrence of food scares and food safety scandals (Ni and Zeng, 2009).

Moreover, confidence in the integrity of the system is low and the problem of fraud is always a threat for an organic food market that is still in its early stages of development. Cases of pesticide-treated produce falsely advertised as organic have been reported in magazines and newspapers: for instance, a top foreign retailer was selling 'organic' foods supplied by a farm that was later found using forbidden agro-chemicals (Tschang, 2007). These matters of fact erode enforceability as well as consumer confidence towards local productions, both of which are essential for a functional organic certification system. Furthermore, since the successful green food initiative was born earlier than the development of an organic movement, competition with the well established green food sector has hindered the further development of the organic market. In fact, in China the green food label is very well known, although it sometimes is confused with the organic label. Organic food and green food are both considered healthy and environment-friendly, but the wider awareness of green food often makes people prefer it to organic food. A linguistic issue may also affect this situation, since in Mandarin language 'organic' is translated in a way that make it sound like 'highly technologic', while 'green' sounds more consistent with the message of safety and environment protection it should deliver (Buckley, 2006).

In April 2005, after 20 years of development, the China National Organic Product Standard (CNOPS) came into force. In combination with 'The Rule on Implementation of Organic Products Certification' the CNOPS defines the scope, normative standards, certification procedure, requirements for certification bodies, use of organic product certification seal, and labelling, as well as importation of organic food products to China. All food products that are sold in China as organic, as well as the food products sold as 'organic in conversion', must comply with the CNOPS. Organic products sold in the Chinese market are required to comply with the following labelling rules:
a. Organic: if the share of certified organic ingredients is higher than 95%.
b. Conversion to organic: if ingredients certified as produced by farm under conversion to organic agriculture constitute more than 95%.
c. Made with organic ingredients: if 70-95% of the final product consists of certified organic or conversion-to-organic ingredients. The percentage of certified ingredients has to be shown on the package.
d. No labelling, but indication in the ingredients table: if the percentage of certified organic or conversion ingredients is less than 70%, the product should not be labelled as organic, but in the ingredient table the organic ingredients may be highlighted (Lu, 2005).

The milestones in the development of the organic food sector in China can be summarized as follows:

- 1984, China University of Agriculture starts research and development regarding eco-agriculture and organic food.
- 1988, Nanjing Environmental Science Research Institute of the State Environment Protection Administration (SEPA) of China starts to study organic food and joins IFOAM in the same year.
- 1990, SEPA participates in the first organic certification inspection in China; the same year the Ministry of Agriculture establishes a China Green Food Development Center to produce 'non-polluted, safe, high-quality nutritious food' for the domestic market.
- 1994, the Organic Food Development Center (OFDC), the first professional body for organic food research, development and certification in China, is established. In the same year, the Organic Food Development Supervision Committee (OFDSC) and Organic Food Certification Committee (OFCC) are established.
- 1995, the OFDC registers the logos of China Organic Food and publishes 'Supervision measures of organic food sign' and 'Technology standards of organic food production and processing'. The Organic Crop Improvement Association (OCIA, a U.S.-accredited certifier) establishes its China Branch and authorizes OFDC as its representative to supervise, determine, and issue certification and to use the organic food signs registered by OCIA.
- 1998, more than 180 different types of farm produce are certified as organic by the OFDC, and are exported all over the world.
- 1999, organic tea is available in Chinese markets.
- 2001, supermarkets in Beijing begin to sell organic food such as fresh vegetables, tea rice, fruit and honey. In August, the Certification and Accreditation Administration of China (CNCA) is established, with its administrative responsibilities authorized by the State Council. It undertakes unified management, supervision and overall coordination of certification and accreditation in China.
- 2002, the China National Accreditation Board for Certifiers (CNAB) is founded by order of CNCA in accordance with relevant laws and regulations, as a national accreditation body responsible for the accreditation of management system certification bodies and product certification bodies. CNAB merges the former China National Accreditation Council for Registrars (CNACR), China National Accreditation Council for Products (CNACP), China National Accreditation Board for Import & Export Enterprises (CNAB) and former China Accreditation Committee for Environmental Management System Certification Bodies (CACEB).
- 2002, the organic market receives a major boost when the China Organic Food Certification Center (COFCC) is created. It is a body of the Ministry of Agriculture.
- 2003, the OFDC gets accreditation from IFOAM. The COFCC is officially established. It is the first organic food certification organ authorized by CNCA. The 'Assessment and Supervision Measures of China Organic Food Production Bases (Trial)' are drawn up. The 'Certification Statute of the People's Republic of China' is enacted and brought into effect.
- 2004, eleven ministries and committees jointly publish suggestions for actively promoting the development of organic food industry. At the end of the year SEPA completed the transfer of the organic food certification work to CNCA.
- 2005, 'Supervision Measures of Organic Food Certification' are officially put into force in China. The CNOPS and the rules on Implementation of Organic Products Certification are issued and are in force. The 'First Forum of China Organic Food Development' and the 'First Congress of China Environmental Protection industry Association Organic food Session', sponsored by SEPA, are held in Beijing. The first organic food import and export association in China is established in Shenyang. Sotore, the first organic food supermarket in China, is

opened in Shanghai. The China Green Food Development Center signs a cooperative Memo of Understanding with German Nuremberg International, Exhibition Company (Biofach) and on November 9-12 an international organic/green food exhibition (Organic Expo) is held in Beijing.

- 2006, the China National Accreditation Service for Conformity Assessment (CNAS) is founded by merging the former CNAB and CNAL.

3. Chinese organic regulations

Organic agriculture and trade is facing a trend of increasing certification and accreditation requirements. According to the State Council Information Office (2007) Good Agricultural Practices (GAP) certification geared to international standards was adopted by 286 export enterprises and agricultural standardization demonstration bases was introduced in 18 pilot provinces; 2,675 food producing enterprises have received HACCP certificates (Xin, 2007).

This is a global trend in developing countries, but in Asia the development seems to be faster: according to Huber (2008). The Asia and Pacific area is the third world's major block in terms of countries with fully implemented organic regulations after Europe, America and the Caribbean (Table 3).

The unique feature of organic regulation in China is that two government departments compete in regulating the question of 'natural food'. The OFDC operates as a certification body for organic food within SEPA of China, while green foods are certified by the CGFDC operating within the Ministry of Agriculture. The fact that Green foods are sometimes also certified as organic may increase the confusion. As mentioned before, in 2005 a detailed regulatory document[8] aimed at regulating the growing market was issued, which outlined guidelines for organic food production, certification, labelling, and sanctions (Baer, 2007).

Certification bodies and product certification bodies, prior to March 31, 2006 referred to the China National Accreditation Board for Certifiers (CNAB) and laboratory to the China National Accreditation Board for Laboratories (CNAL); now, according to the China national organic product standard, the China National Accreditation Service for Conformity Assessment (CNAS) has replaced both CNAB and CNAL. All certification bodies have to be approved by CNCA

Table 3. Organic regulations by continent (Huber, 2008).

	Countries with regulations	Countries per continent	Percentage
Europe	39	41	95
America and Caribbean	17	35	49
Asia and Pacific	11	62	18
Africa	3	55	5
Total	55	193	33

---

[8] *Administrative measures for certification of organic products* promulgated by China's Council of General Administration on Quality Supervision, Inspection, and Quarantine, AQSIQ Decree No. 67 2004.

and must be accredited by CNAS according to international standard ISO 65. Certified organic products must be labelled with both the China national organic logo and the certifier's logo. The certification bodies must employ inspectors who are registered with the China National Auditor and Training Accreditation Board (CNAT). Figure 2 shows the logos of the Agencies currently operating in the Chinese organic certification system.

The supervision of the certification bodies and the administration of organic certification at the local level are under the responsibility of the General Administration of Quality Supervision, Inspection and Quarantine of China (AQSIQ). Figure 3 shows the organization of standards and conformity assessment bodies in China.

3.1 Chinese organic certification bodies

Seventy-four countries have a home-based certification organisation. Most certification bodies are located in developed countries and also offer their certification services in developing countries. Asia has 147 certification bodies (Table 4), most of them based in South Korea, China, India, and Japan; large parts of the continent still lack local service providers.

As mentioned above, a certification body is allowed to engage in organic certification activities only after it has been approved by the CNCA and acquires legal person's status according to law. The minimum requirements are:
a. fixed premises and necessary facilities;
b. management systems that meets the requirements for certification and accreditation;
c. a registered capital higher than 3 million RMB;
d. at least 10 full-time certification personnel in relevant fields.

Certification bodies should get accreditation from CNAS and organic inspectors have to get registration at the CNAT after formal inspector training and at least 15-days inspection practices. CNAT has developed an Organic inspector training manual for this and accredited 3 training organizations to conduct the training. It has also developed a set of rules for inspector registration.

The standards organic products BG/T 19630-2005 were issued on January 1, 2005 and became effective on April 4, 2005. They consist of 4 sub standards (Part 1 Production GB/T 19630.1-2005, Part 2 Processing GB/T 19630.2-2005-4-22, Part 3 Labelling and marketing GB/T 19630.3-2005, and Part 4 Management system BG/T 19630.4-2005). The four parts can be used as a whole

China National Accreditation
Service for Conformity
Assessment (CNAS)

China National Auditor and
Training Accreditation Board
(CNAT)

Certification and Accreditation
Administration of China
(CNCA)

Figure 2. Certification, supervision and administration agencies' logos.

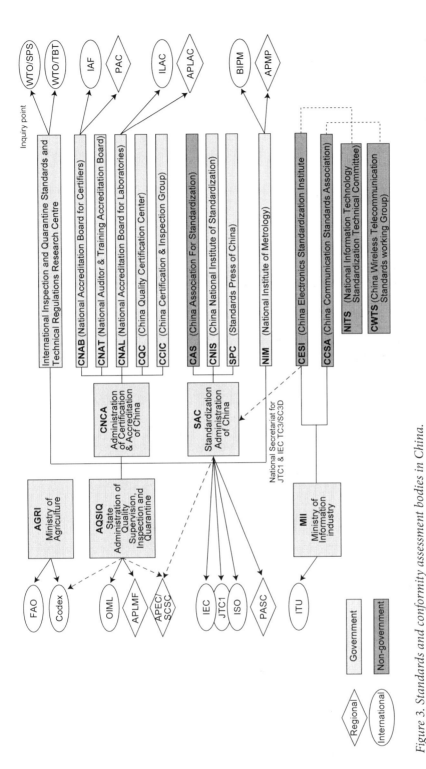

Figure 3. Standards and conformity assessment bodies in China.
FAO: Food and Agriculture Organization, OIML: International Organization of Legal Metrology APLMF: Asia Pacific Legal Metrology Forum, APEC: Asia-Pacific Economic Cooperation, IEC: International Electrotechnical Commission, JTC1: Joint Technical Committee 1, ISO: International Organization for Standardization, PASC: Pacific Area Standards Congress, ITU: International Telecommunication Union, WTO: World Trade Organization, SPS: Information Management System, TBT: Technical Barriers to Trade, IAF: International Accreditation Forum, PAC: Pacific Accreditation Cooperation, ILAC: International Laboratory Accreditation Cooperation, APLAC: Asia Pacific Laboratory Accreditation Cooperation, BIPM: Bureau International des Poids et Mesures, APMP: Association of Proposal Management Professionals.

Table 4. Number of certification bodies per continent (Huber, 2008).

	2007	2006	2005	2004	2003
Africa	8	8	7	9	7
Asia	147	93	117	91	83
Europe	172	160	157	142	130
Latin America & Caribbean	47	43	43	33	33
North America	83	80	84	97	101
Oceania	11	11	11	11	10
Total	468	395	420	383	364

system, but also separately for different activities. Its roles are production and processing as well as certification. The standards were developed based on the principles and requirements of IFOAM basic standards for organic production and processing. Points from the Codex Alimentarius, the former EU Organic Regulation 2092/1991 (and subsequent regulations completing and updating the EU standards), NOP (US National Organic Programme), etc, were also considered in this. In its content, China's organic standards are compatible with those standards applied for the purposes of standard harmonization internationally and for the promotion of world organic trade. The fields that the standards cover are not only crop and animal production, but also aquatic production, in which water-animal feeding and open-area fishing are included. Furthermore, it covers edible products and inedible products, such as organic fibre and textiles. Therefore, the standards cover all organic fields at present. In contrast to organic standards in other parts of the world, Chinese standards add a specific element regarding a management system for improving the management of producers, processors and handlers. It introduced the principles and methods from ISO (International Organization for Standardization) 9001:2000 quality management system to organic management, which specifies file and record management, resource management, internal inspection, audit trail system, and continuous improvement (Lu, 2005).

Chinese organic certification bodies are also allowed to propose and introduce their own, more strict, private standards for organic agriculture and to certify products accordingly. provided that they also comply with national regulations and that the private standards and certification procedures are made public and freely available. Organic products, therefore, might bear both the national 'organic products seal' and the logo of the private standard issued by the certification body.

According to a recent CNCA report (Shi, 2007), 27 organic certifiers are approved by CNCA, but the major 4 certifiers, which are COFCC, OFDC-China, WIT Assessment and OTRDC (Organic Tea R&D Center of Tea Research Institute under the Chinese Academy of Agricultural Sciences) account for over 80% of organic product certification in China. However, the vast majority of organic food certification in China is issued by foreign bodies, especially from the EU.

3.2 Major foreign organic food certification bodies in China

Organic food production in China is still mostly export-oriented, but the national certification bodies are not able to offer internationally recognized certification, since the equivalence of the Chinese regulations is not yet recognized by the public authorities in the most important export markets. Therefore, organic products are frequently certified directly by foreign private certification bodies, which usually establish certification offices in China, setting up joint ventures with local partners. Local inspectors and foreign certification bodies jointly implement the auditing procedures and perform inspection activities in order to ensure that both compliance with international rules and trust and collaboration with organic farmers and food producers are assured. Many certifiers are competing together for a rather limited potential market of certifiable producers. However, this market is not easily accessible to local certifiers since, in most cases, the foreign trader decides which certification label should be used on his products.

In 2005 the OFDC signed a mutual recognition agreement with 31 foreign certification bodies that are accredited by the IFOAM, which provides applicants with great convenience in the mutual recognition of certificates among these organic food certification bodies.

Foreign organic certifiers, however, may face several difficulties in performing their control activities. In fact, the agriculture and food production system in China is quite different from those in the EU or US, and international organic inspectors may not be able to understand its specificities, which are heavily influenced by politics and by the peculiar social structure. For instance, the situation documented by Paul Thiers of Washington State University about the challenges and unusual problems faced by foreign inspectors, such as different cultural approaches toward concepts like 'conflict of interest', 'deception', 'manipulation' and 'denial of access'. Quoting Paul Thiers, 'Inspectors who are unclear if they are in a village or a township, who do not anticipate the revocation of land tenure contracts, or who do not understand the difference between a mayor and a party secretary are at a distinct disadvantage in anticipating the effects of self-interest and institutional structure on behaviour' (Blobaum, 2007).

4. Organic food market

China's organic food farming amounts to only 0.1% of the total agriculture production and represents only 0.76% of the total arable land in China; by comparison, organic food has a share of from 3% to 5% of the overall value of agricultural products on the world food market, and organic arable land covers 5 to 10% of the world arable land (Yang, 2007). Organic agriculture and food production started as an export-only activity since in the mid-1990s the domestic market for organic food was virtually inexistent (Buckley, 2006). The domestic market share of organic food is now less than 0.1%, much lower than the average level of 2% in the world; nevertheless, although small the market is emerging. The development of a domestic organic food market has been mainly driven by the emergence of food scares and thanks to the interest of the growing middle class, which tends to view organic products as safer and healthier (Luis and Firmino, 2007). A sustained relevant growth of this still tiny market niche in continental China may also be anticipated considering the growing interest for organic food that has been reported in the Taiwanese market, which is culturally very similar (Chen, 2007).

On the supply side, stores in China's main cities are currently offering organic alternatives in several food categories, in response to an increasing demand for safer products (Baer, 2007). The exposure of China's food safety weaknesses, which decreases the already low level of confidence in the Chinese system, the rapid growth of demand of organic food and the slow adaptation of domestic supply, combine to create the phenomenon by which some organic food initially sold to international food suppliers and distributors returns to China as westernized, higher-quality, and expensive goods. As an example of this situation, a 50% yearly increase in organic food sales, mainly driven by imported food, was claimed by Carrefour in China in 2007 (Rein, 2007).

These trends suggest that a growing demand for safe foods is to be expected and new opportunities in China's domestic market, as well as a growth in intra-Asian trade may boost a further substantial development of organic farming in this country. The lack of harmonization of standards between countries, however, 'adds to the burden for farmers and traders who must select and sometimes use multiple certifiers, and naturally, it hinders their access to international markets' (IFAD, 2005).

According to OFDC, in fact, the domestic market is also on the move, thanks to the growing demand for environment-friendly products. The value of domestic sales currently amounts to around 40 billion RMB, as shown in Table 5.

4.1 Price

Organic certification provides produce and food products with a distinctive feature that helps farmers and food producers to differentiate and to avoid fierce competition in the conventional market. The opportunities to capture higher value for the organic process mostly depend on the recognition of higher quality and on the willingness on the part of consumers to pay a higher price. Markets that recognize the superior quality of organic products and that show a demand adequate to the level of supply pay a premium (Giovannucci, 2006). Organic products can also be marketed as conventional products, in cases where the market for organic products has not yet developed. This is often the case in several developing countries where organic products are mainly produced for the export market. In recent years, however, also in developing countries the domestic demand for organic products has grown remarkably. The main drivers of this phenomenon have been the growth of household income and the public's increased concern regarding health issues and food safety. The Chinese domestic market offers considerable opportunities because of the country's population size and since an increasing proportion of households may dispose of a significant part of their income. These households are mostly located in urban areas, where the growth of food availability and sales do not depend anymore on food supplies from the surrounding rural areas. Most of the organic food retailers, in fact, are located in the urban areas.

According to a recent research (Wei and Yinchu, 2007), it emerged that the main reasons why Chinese consumers buy organic vegetables are that they consider them safer (42.9%), healthier and more nutritious (27%) and environmentally friendly (2.5%). Apparently, these are favourable conditions for a rapid growth in demand, since health, nutrition and the environment are hot topics among the Chinese public.

However, the domestic market for organic food is still rather small, although the more affluent consumers now have the disposable income to be able to afford to buy organic food. In fact, 19% of the Chinese population, almost 250 million people, earn more than 20,000 euro per year, which is a substantial income in China. According to IFAD (2005), the fact that the demand for organic food is still modest in size is mainly due to:

Table 5. Organic and green food production and Market Size (2005) (OFDC, http://www.ofdc.org.cn).

	Green food	Organic food
Participating corporations	3,695	416
Items of products	9,728	1,249
Products weights (million tons)	6,300	66.9
Sales value (billion RMB)	1,030	37.1
Export value (billion US$)	16.2	1.36

- Modest availability and selection in stores exacerbated by limited prominence.
- Inconsistent supply from farmers.
- Sometimes exorbitant prices. Organic food products in China can cost up to 12 times more than the same products from conventional agriculture (e.g. organic green onions cost 19 RMB/kg, while organic tomatoes cost 17 RMB/kg).
- Poor consumer understanding of organics. Currently, the main obstacle to further development of China's organic sector is public awareness: 'many people want to buy healthy food, but they don't know where to find it or how to buy it, and many farmers have organic food, but don't know how to market it' (Yang *et al.*, 2006).

4.2 China's organic exports

Due to the high price and scarce awareness among Chinese consumers, organic food is mostly produced for export purposes. Most Chinese organic food is exported to developed countries such as Japan, USA, Canada and the EU. Before 1998, almost all organic food certification bodies came from developed countries, most of these were based in the EU, especially Germany. China's organic food exports represent less than 1% of the world organic food supply. Tea, beans and rice are the major organic export products, which are mainly produced in China's coastal regions (Yang, 2007). Organic food export volumes and the range of the products are both limited, but growing. Between 2003 and 2005, China's organic food exports grew from 142 million US$ to 350 million US$ (Bezlova, 2006). Most of the organic food is raw farm produce. Unlike western countries, the decision to adopt organic farming is usually dependent on government policy rather than on farmers' decision (Baer, 2007).

In China, over 60% of the provinces and cities are beginning to produce organic food, mostly in the northern, middle and western parts of China. However, the distribution of organic food production facilities is rather imbalanced, for there is no organic food industry in certain provinces and cities.

5. Conclusions

Currently, China is in the difficult position of having to address environmental damage caused by past misuse of chemicals in agriculture, and feed a population that is growing in number and demanding more and better quality food. Organic farming could both reduce pollution in rural areas and bring higher income to Chinese farmers, and switching to sustainable agriculture would not be complicated since China's traditional agricultural cultivation techniques which are highly suitable for organic production are still currently practised in many rural regions of China (Sheng *et al.*, 2009).

Unfortunately, the certification process is demanding and expensive. A debate is still under way with regard to allowing farmers to apply for cooperative-level certification, adopting a sort of participating certification scheme, meaning that the auditing process fully involves all the actors and stakeholders (farmers, consumers, etc.) who are involved in the food chain. This would probably allow the level of confidence among the actors to increase, reducing transaction costs and the financial burden for the organic supply chain, therefore opening new opportunities for the domestic market.

Although the domestic market for organic food is growing, it is still underdeveloped and remains a tiny niche market. Exports are the major reason for growth, with only limited distribution nationwide, while green food satisfies domestic demand for higher quality products. Therefore, green food is the strongest competitor for organic food in the domestic market.

Chinese producers mostly supply organic ingredients, which command comparatively low prices compared to international traders and processors. This does not create incentives for establishing sustainable production systems.

The lack of specialized technical knowhow about organic production technologies further hinders the possibility of converting farms to organic. An active farmer learning-support system, through stronger links with higher education and research institutions or through the creation of professional support services aimed at guiding operators through norms and standards, could help solve the problem, but in any case a precondition would be the full support of the government, both in terms of policies and incentives.

However, the main impediment to further development of China's organic sector remains public awareness: people are unaware of the difference between pollution-free, green and organic food, and distrust local certification due to law enforcement's bad reputation (Ni and Zeng, 2009).

Finally, Chinese integration into international organic food trade regimes raises challenges for the harmonization of standards and certification procedures, which are not yet fully compatible. Establishing a legal framework aimed at implementing the legal guidelines for organic standards and enforcing existing regulations (the country has over 200 individual food safety laws, regulations and standards, which apply at national and regional level, according to official media) would be one of the preconditions for the recognition of equivalence of organic products from China by the importing countries.

Discussion in lecture room

Apply your theoretical knowledge with respect to this paper:
- Read the paper *Factors influencing purchasing decisions of Austrian distribution channel operators towards 'made in China' organic foods* by Haas *et al.* (2010). Where do you see the future potential for Chinese organic food worldwide?
- Read the paper *Role of certification bodies in the organic production system* by Canavari *et al.* (2010) about the role of certification bodies in the organic production system and discuss their results in respect to the Chinese situation.
- Search for scientific articles about the potential of organic agriculture to feed the world population. Do you see organic agriculture as a niche alternative for the Chinese agriculture, or could it become a form of agriculture overtaking green food production in the long run?

References

Baer, N., 2007. The Spread of Organic Food in China. A China Environmental Health Project Research Brief - China Environment Forum, The Woodrow Wilson International Center for Scholars, Washington, DC, USA.

Bezlova, A., 2006. China Experiencing Organic Farming Boom. DAWN Group of Newspapers, available at: http://www.dawn.com/2006/06/25/int12.htm. Accessed July 2009.

Blobaum, R., 2007. Inside Organics: Surprise NOP Auditor Visits to Organic Farms and Processors in China is Overdue Response to Concerns About Integrity of Organic Food Imports. The organic broadcaster, Midwest Organic and Sustainable Education Service (MOSES), available at: http://www.mosesorganic.org/attachments/broadcaster/roger15.5china.html. Accessed July 2009.

Buckley, L., 2006. Pathbreaking Newsletter Promotes Development of Organic Sector in China, Worldwatch Institute, Washington, DC, USA, available at: http://www.worldwatch.org/node/3887. Accessed July 2009.

Canavari, M., Cantore, N., Pignatti, E. and R. Spadoni, 2010. Role of certification bodies in the organic production system. In: R. Haas, M., Canavari, B. Slee, C. Tong and B. Anurugsa (eds.), Looking east looking west: organic and quality food marketing in Asia and Europe. Wageningen Academic Publishers, Wageningen, the Netherlands, pp. 85-99.

Chen, M.F., 2007. Consumer attitudes and purchase intentions in relation to organic foods in Taiwan: moderating effects of food-related personality traits. Food Quality and Preference 18: 1008-1021.

Giovannucci, D., 2006. Salient Trends in Organic Standards: Opportunities and Challenges for Developing Countries. Paper prepared for the World Bank/USAID Trade and Standards E-learning Course, Jan-March, 2006. Available at: SSRN: http://ssrn.com/abstract=996093. Accessed July 2009.

Haas, R., Ameseder, C. and Liu, R., 2010. Factors influencing purchasing decisions of Austrian distribution channel operators towards 'made in China' organic foods. In: R. Haas, M., Canavari, B. Slee, C. Tong and B. Anurugsa (eds.), Looking east looking west: organic and quality food marketing in Asia and Europe. Wageningen Academic Publishers, Wageningen, the Netherlands, pp. 115-126.

Huber, B., 2008. The World of Organic Agriculture: Regulations and Certification: Emerging Trends 2008. BioFach Congress 2008, Nuremberg, Germany.

IFAD, 2005. Overview of markets and marketing. In: Organic Agriculture and Poverty Reduction in Asia: China and India Focus. International Fund for Agricultural Development (IFAD), Roma, Italy, available at: http://www.ifad.org/evaluation/public_html/eksyst/doc/thematic/organic/asia.pdf. Accessed July 2009.

Lampkin, N., Foster, C., Padel, S., and Midmore, P., 1999. The Policy and Regulatory Environment for Organic Farming in Europe. In Organic Farming in Europe: Economics and Policy (Vol. 2). Hohenheim University, Stuttgart, Germany.

Lu, Z., 2005. Organic certification and accreditation in China. Paper presented at the 15th IFOAM Organic World Congress, September 20-23, 2005, Adelaide, Australia.

Lu, Z., 2002. Production and Market of Organic Foods in China. Proceedings of the 1st RDA/ARNOA International Conference Development of Basic Standard for Organic Rice Cultivation, 12-15 November 2002, RDA and Dankook University, Suwon, Korea. Available at: http://www2.rda.go.kr/kpms/ipsm/Korean/03_undp/.../OA-11(Lu).doc. Accessed January 2009.

Luis, E., and Firmino, A., 2007. Organic Agriculture in China and Viet Nam: Nanjing and Hanoi as Case Studies. The Hague: Agricultural Economics Research Institute, SEARUSYN Project. Available at: http://www.searusyn.org/files/fc05a06a09f6c7e0e1812cbc4ebf401d.pdf. Accessed January 2009.

Ni, H.-G., and Zeng H., 2009. Law enforcement is key to China's food safety. Environmental Pollution 157: 1990-1992.

OCA, 2003. Pesticide Residues a Major Threat to China's Ag Exports. Organic Consumers Association, Finland, MN, USA, available at: http://www.organicconsumers.org/toxic/chinapesticides012103.cfm. Accessed January 2009.

Paull, J., 2007. China's Organic Revolution. Journal of Organic Systems 2: 1-11.

Paull, J., 2008a. The Greening of China's Food - Green Food, Organic Food, and Eco-labelling. Sustainable Consumption and Alternative Agri-Food Systems Conference. Liege University, Arlon, Belgium, available at: http://www.orgprints.org/13563. Accessed January 2009.

Paull, J., 2008b. Green Food in China. Elementals. Journal of Bio-Dynamics Tasmania 91: 48-53.

Rein, S., 2007. China Seeks Quality from Multinationals. Forbes.com, available at: http://www.forbes.com/opinions/2007/10/03/china-quality-retailing-oped-cx_shr_1003china.html. Accessed January 2009.

SCIO, 2007. The Quality and Safety of Food in China. State Council Information Office, Beijing, available at: http://www.china.org.cn/english/features/book/221926.htm. Accessed January 2009.

Sheng, J., Shen, L., Qiao, Y., Yu, M., and Fan, B., 2009. Market trends and accreditation systems for organic food in China. Trends in Food Science and Technology 20: 396-401.

Shi X., 2007. Progress Report on Organic Certification in China. In: Updates from ITF Members and Country Reports. 7th ITF Meeting, November 2007, International Task Force on Harmonization and Equivalence in Organic Agriculture, FAO-IFOAM-UNCTAD, Bali, Indonesia.

Tschang, C.-C., 2007. Organic, With Pesticides (Extended). Business Week. Available at: http://www.businessweek.com/magazine/content/07_31/b4044062.htm. Accessed July 2009.

Veeck, A., and Veeck, G., 2000. Consumer Segmentation and Changing Food Purchase Patterns in Nanjing, PRC. World Development 28: 457-471.

Wei, X., and Yinchu, Z., 2007. Consumer's Willingness to Pay for Organic Food in the Perspective of Meta-analysis, WERA 101 Annual Conference. Renmin University of China, Shanghai, China.

WHO, 2007. Environment and health in China today. World Health Organization Representative Office in China, available at: http://www.wpro.who.int/china/sites/ehe/overview.htm. Accessed January 2009.

Willer, H., and Kilcher, L. (eds.), 2009. The World of Organic Agriculture. Statistics and Emerging Trends 2009. FIBL-IFOAM Report. IFOAM, Bonn; FiBL, Frick; ITC, Geneva.

Xia, W., and Zeng, Y., 2007. Consumer's attitudes and willingness-to-pay for Green food in Beijing. School of Agricultural Economics and Rural Development, Renmin University of China, China.

Xin, H., 2007. White Paper on Food Quality and Safety. China Daily. Available at: http://www.chinadaily.com.cn/china/2007-08/17/content_6032557.htm. Accessed July 2009.

Yang, M., Jewison, M., and Greene, C., 2006. Organic Products Market in China. GAIN Report Number: CH6405. USDA Foreign Agricultural Service, Washington, DC, USA.

Yang, Y., 2007. Pesticides and Environmental Health Trends in China. China Environment Forum, Woodrow Wilson International Center for Scholars, Washington, DC, USA.

Zhang, X., 2005. Chinese Consumers' Concerns About Food Safety: Case of Tianjin. Journal of International Food & Agribusiness Marketing 17: 57-69.

Market access for smallholder livestock producers under globalised food trade: an analysis of contract farming and other market-linking institutions in India, Thailand, Vietnam and the Philippines[9]

A. Costales[1] and M.A.O. Catelo[2]
[1]Pro-Poor Livestock Policy Initiative (PPLPI), Food and Agriculture Organization of the U.N., Animal Production and Health Division, Viale delle Terme di Caracalla, 00153 Rome, Italy
[2]University of the Philippines Los Baños, Department of Economics, College of Economics and Management, Laguna, Philippines

Abstract

Under globalised food trade, agricultural production processes and marketing have become tightly linked along supply chains, making it more difficult for smallholder producers to remain integrated in mainstream markets as transactions costs of doing so increase. This paper provides an assessment of formal and informal contracts in efficiently integrating rural smallholder producers to mainstream markets of livestock products and looks into four country case studies. It finds that studies on contract farming have focused mainly on formal contracts that tended toward larger-scale farmers in peri-urban locations. Concerted efforts toward informal contracts and alternative directions would result in the opening up of more windows for smallholders that respond to evolving demand patterns for livestock products in domestic markets, rather than forcing the issue of enabling them to produce the high-end products for export and supermarkets whose tight product and process standards are extremely difficult for them to hurdle.

1. Background

Agricultural and food markets have, in the past 20 years, dramatically changed to become more integrated, globalised, and consumer driven. In Asia and Latin America, these changes are leading to rapid commercialization of agriculture. Within the context of a commercializing and globalizing food system, there is now a much higher degree of integration between producers and the output market where product standards related to quality and food safety are given relatively greater importance. Pingali, *et al.* (2005) argue that agricultural production processes, except for purely subsistence farming systems, are now increasingly linked via supply chains, to markets for inputs, to enterprises undertaking the processing of output, to those distributing the final products, and to consumers demanding particular attributes of farm products.

The intensified commercialization of agriculture is transforming traditional food systems from the simple, 'essentially production systems' that employed crude methods for food processing and involved minimal distribution channels, to ones that are now modern, complex and even quite discriminatory with respect to putting up participation barriers against potential entrants. The emergence of modern agricultural food systems has brought forth more complex relationships and links within the supply chain and these relationships also occur under particular market infrastructural and institutional environments. Thus, larger amounts of costly information

---

[9] This paper is derived from the original FAO PPLPI Final Report on Contract Farming and other Market-linking Institutions as Mechanism for Integrating Rural Smallholder Livestock Producers in the Growth and Development of the Livestock Sector in Developing Countries, submitted by the same authors in May 2008.

and knowledge (e.g. technological, managerial, monitoring, etc.) have to be disseminated and processed within and between each link of the supply chain to minimize informational uncertainties within the food system. New physical (asset) investments have to be undertaken in order to meet technological requirements associated with compliance to increasing quality and safety standards. Moreover, as the level of agricultural commercialization increases, the need for highly specialized production units mirrors the need for tighter control and supervision along the supply chain. By implication, modern food systems in the highly commercialized agricultural markets have introduced a new set of transaction costs that raised the cost of entry into certain products as well (Reardon and Timmer, 2005; Pingali, *et al.*, 2005; Pingali, 2006). Market failure in any one, or more, of the links also has implications on the ability of producers to transcend barriers to access to markets for inputs and/or products, and critical services (Minot, 1986; Rich and Narrod, 2005).

Among the main drivers of the transformation taking place in agriculture in developing countries are the changing demand conditions brought about by growth in population, increases in real per capita incomes, income elasticity of demand, urbanization, and variations in real prices. Countries are also experiencing a change in tastes and preferences of their population for food products toward those considered as 'superior' (Baumann, 2000; Da Silva, 2005; Delgado, *et al.*, 1999, 2008; Kirsten and Sartorius, 2002; Simmons, 2002; Tiongco, *et al.*, 2006).

Diets are also changing and there is increasing demand for higher-value products over staples, product quality, food safety, and convenience. With a greater number of women in the labor force, demand for processed products and pre-prepared foods is also higher (Pingali *et al.*, 2005). Consumers have also become increasingly discriminating (Babcock, 2007; Delgado *et al.*, 2003; Kirsten and Sartorius, 2002; Paulson and Costales *et al.*, 2007). Public awareness on eating the right foods, health risks of consuming unsafe food, and environmental and animal welfare implications of production processes is growing (Da Silva, 2005; Tiongco, *et al.*, 2007). Furthermore, there is evidence of increasing disposition of consumers toward eating meals outside the home. Consumers are willing to pay a 'price premium' for quality, freshness and safety in food products. Expectedly, these food safety issues impact more on fresh, perishable food products such as fruits and vegetables, fresh meat and seafood. These are major export products of a number of developing countries. Corollary to this, consumers now demand information on labels of food products.

Given the foregoing structural changes, agri-food systems, therefore, needed to adapt to changing consumer preferences, inducing them to have tighter control and coordination over production and handling processes such as cold storage refrigeration, and transport to ensure the steady and timely supply of safe food products. Thus, we witness the 'industrialization' or the birth of 'new agriculture' (Da Silva, 2005; Kirsten and Sartorius, 2002; Sartorius and Kirsten, 2007) where traditionally small and family-based farms are replaced by commercial or intensive factory farming. Agri-food systems are also transformed into vertically coordinated modes of governance along the supply chain, involving producers, integrators/processors, wholesalers and retailers, and even supermarkets.

To remain competitive in such a dynamic agricultural market environment, firms must find alternative governance structures to exercise greater control in the production, processing and distribution functions. Firms accomplish this by either fully vertically integrating, or by engaging in contracts with farmers, engaging in strategic alliances (partnerships), or engaging in a merger with the enterprise with which it had previously been engaged in a market transaction (Birthal *et al.*, 2006; Catelo and Costales, 2008; Fairoze *et al.*, 2006; Peterson and Wysocki, 1997; Son *et al.*, 2008).

Within this context, contract farming has, in recent years, been presented as a potentially effective market-oriented institution in reducing transaction costs and in bridging the gap between the rural smallholder producer's resources, assets, and capacities on the one hand, and the increasingly stringent demands of the consumers on the other. The assessment of contract farming as an instument in efficiently and profitably integrating rural smallholder producers to mainstream markets of meat and milk products is the purpose of this review.

2. Materials and methods

This work relied on an extensive review of the theoretical literature on contracts and transaction costs economics, and empirical literature on contract farming in developing countries in agriculture in general, and in livestock products in particular. More detailed reviews were also undertaken on case studies in contract farming in India, Thailand, the Philippines, and Vietnam, in the production and marketing of milk, broiler chicken, and pigs.

3. Results and discussion

For rural smallholders to perform at par with better-equipped farmers and participate effectively in these supply chains of high-value products, they must have access to the following: (1) intelligence (information) on these markets (products, consumers, location, volume, required characteristics); (2) knowledge and skill in the technology to produce particular products; (3) required facilities, equipment, and reliable supply of specific inputs, and requisite financial services; and (4) institutions that provide mechanisms to guarantee or certify that the products of smallholders do possess the desired characteristics, particularly when such characteristics are not directly and immediately observable. Absent or incomplete access due to particular market failures or/and public goods provision failures, the configurations in the supply chains and the accompanying developments in product and process standards work to impose barriers on rural smallholders, and exclude them from access to the very markets in which the demand for meat and milk products are rapidly expanding. The public sector in developing countries is in a limited position to attend to the specific needs of smallholder producers in dynamically changing markets for livestock products, given the constraints to public budgets and the capacities of government personnel. Within this context, the literature has looked at contract farming as a market-oriented institution that can link smallholder producers with consumers with demand for differentiated products and product characteristics.

Contract farming generally refers to a binding arrangement between two parties consisting of an agro-processing firm (contractor) and an individual producer (contractee) in which they engage in 'forward agreements', with well-defined obligations and remuneration for tasks done, often with specifications on product properties such as volume, quality, and timing of delivery. This arrangement basically permits the firm to exercise influence on production processes that are delegated to independent farms in a manner that is consistent with its objectives. Contract farming belongs to the set of 'hybrid' transaction coordination schemes between spot markets on one side (pure 'invisible hand' coordination through market exchange) and an extreme and full vertical integration ('hierarchical' decision making) on the other, where the choice of transaction organization is influenced by the objective of reducing transaction costs, among others.

In general, most smallholders do not participate in contracts but independently produce and sell in spot markets. The review revealed that studies on contract farming have focused mainly on formal contracts. In the case studies, formal contracts were written and signed contracts between an integrator company (livestock breeder, feed processor, meat or milk processor, packer and distributor) and a farmer. They were often adaptations from formal contract farming agreements in developed country settings. Informal contracts, on the other hand, were more diverse in form and in terms of contract, ranging from agreements within a group of voluntary farmers' organization or cooperative, sub-contractors of another market intermediary with a formal contract to supply a particular product to an integrator company, or an agreement between a farmer and a trader on the guaranteed supply of inputs and/or the guaranteed delivery of a product to the trader. Consistent with earlier reviews, in terms of the ability of contract farming in efficiently and profitably integrating rural smallholder producers in high-value markets, the review revealed rather mixed results, where there were some promising successful cases, and many failed ones.

In the case studies in the branch of broiler chickens, formal contracts were more prevalent than informal contracts. In contrast, in pig production in the Philippines and Vietnam, informal contracts dominated over formal contracts. In milk production, informal contracts are more prevalent than formal ones in India where the final products are fresh milk and traditional dairy products, but the reverse is the case in Thailand and Vietnam, where the final product is ultra-high temperature (UHT) processed milk.

Formal contracts rather tended toward the larger-scale farmers in peri-urban locations than towards rural smallholders. They usually required a form of a 'bond' as collateral to secure the integrator company's initial exposure to the contract grower. Products involved were generally industrial-type livestock, requiring specific types of industrial inputs (e.g. day-old-chicks of exotic breed, formula mixed feeds). Destination markets are major urban centres, through formal market chains. The terms in the contract documents are detailed and specific, stipulating obligations of each party, the production processes and protocols that need to be followed, the productivity standards that need to be met, and the corresponding incentives (penalties) that are given when standards are exceeded (not reached), such as with feed conversion ratios (FCR) and mortality rates. Producer remuneration was dominantly of the wage contract ('fixed fee') form, where the farmer gets paid a fixed amount for each unit weight or volume of output produced, although there are a few instances where the terms of the contract reflected 'guaranteed forward price' or 'profit-sharing' agreements. Farmer performance was closely monitored by integrator company personnel. Contracts are enforced internally where poor performance is penalised by lower remuneration, and violations of terms of contract are penalised with non-renewal of contract, apart from other measures.

Formal contracts in developing countries often reflected a relationship between an individual farmer with less information and social capital and a dominant large integrator company with more information and clout, where monopsonistic or oligopsonistic market powers could be exerted to tilt the distribution of benefits from the relationship to the advantage of the integrator company. These could be tempered, however, by better information by the contract farmers, better organization, as well as a stronger institutional environment that offers more equitable protection between parties of a formal contract.

The tendency of formal contract to favour larger farmers stems from the economies of scale achieved by integrator companies in dealing with fewer suppliers with larger volumes to offer, as well as the reduction of transaction costs in dealing with and monitoring various smallholders with different capacities to deliver. Smallholders who supply output only intermittently were perceived to be more prone to reneging on contracts than larger producers when the prices offered by the spot market become more attractive than the contracted price.

There were, however, exceptions in the association between formal contracts and larger farmers, with formal contracts found in pig production involving a feedmilling cooperative and smallholder producers who were members of the cooperative. The indications are that smallholders are included in formal contracts when small farms are the dominant production system and suppliers in locations where the integrator company operates, when smallholders possess the human capital and are receptive of training within the system, or/and when the integration of smallholders in a particular location in the supply chain is an explicit goal of the integrator company.

Informal contracts, on the other hand, have been found to be more inclusive of smallholders, although informal contracts also covered farmers with larger scales of production. Entry into such contracts involved prior social capital (e.g. membership in a farmers' organization, earned reputation) rather than the demonstration of physical collateral. Product types varied, where industrial or semi-industrial-type pigs were produced, but transactions with indigenous pigs were also accommodated. In general, destination markets were domestic urban markets, through informal market chains. However, output of smallholder sub-contractors also needed to pass quality requirements when the market intermediary was linked with an integrator company through a formal contract. Greater flexibility characterized the terms and the enforcement of agreements in informal contracts. Although explicit benefits from higher output prices or lower input prices were not always exhibited, other more 'implicit' benefits such as assurance of inputs, delivery services, and greater assurance of output markets were highly valued. The sustainability of the transactions was determined by the firming up of reputation and trustworthiness in delivering one's part of the agreement in repeated successful transactions. Thus, there was also a selection process that evolved in informal contracts, with those falling out relegated to deal with the spot markets in the selling of products and the purchase of inputs.

4. Conclusions and areas for policy intervention

Purpose and effect of contract farming is to link consumers with particular demands on products and product characteristics more closely with producers. These producers serve as market-mediating institutions, which organize the production of such outputs with farmers-on-contract and the processing and packaging of the same to identified distribution centres. There is no imperative, however, that contracts be formal, unless the destination of the products were export markets or high-end domestic supermarkets that require the most rigorous standards, where formal identity is essential for traceability. Such contracts, by the nature of their requirements, understandably, are biased toward farmers with highly specialized assets and skills, where inclusion or exclusion is determined by a very slim margin of error. Such types of contracts are beyond the reach of the typical rural smallholder in developing countries.

The area of high-value livestock products for exports and for supermarkets has traditionally not been the realm of the vast majority of smallholders. This is not expected to radically change in the near future gauging from the distance between sophisticated capacities needed to meet formal standards for product quality and food safety and the capacity of smallholder livestock producers and the standards prevailing in informal market institutions which they rely upon to engage in trade with their products. On the other hand, there is that larger segment of domestic market demand for livestock products by significant sections of households in main urban centres as well as smaller towns, toward which most smallholder production is directed. For these consumers, demand for differentiated products is also expressed in terms of their willingness to pay a price premium to obtain livestock products with distinct quality characteristics (freshness, taste, flavour, texture, among others) that normally are associated with non-industrial-type of livestock breeds or production processes. The growth of such domestic markets for differentiated livestock products, and the development of alternative institutional market arrangements that would more efficiently communicate information on product differentiation (quality, production process, food safety), and a system to guarantee credibility of such qualitative differences, should work to create value where it is sought. This should provide incentives for smallholders to create greater value in the livestock activities that they currently undertake. At the same time, the development of such domestic market would also offer consumers of all income brackets a wider range of livestock products to choose from, not just among branded products distributed in supermarkets.

Informal contracts have been accorded scant attention in the investigation of contract farming as a mechanism of improving market access of producers of livestock products in developing countries. This has mainly been due to the focus on strengthening market access to exports and supermarkets in the realm of high-value products, where products are standard industrial-type meat and milk products, with stringent product and process standards. Export markets, however, constitute only a small share in the market for smallholder meat and milk products output. Even in domestic markets of developing countries, in the realm of fresh meat and milk products in general, the market share of supermarkets continues to be significantly smaller than that of the open or other more informal outlets. It still remains to be empirically established whether in the next decade, the supermarket share will soon outstrip that of its traditional informal counterpart. It is not inevitable that as developing countries economically improve, product distribution will shift linearly from open markets to supermarkets. Even among highly urbanized populations in relatively developed countries in Asia, people who go to supermarkets also diversify and go to traditional retail outlets to obtain products of particular characteristics or derive particular services not obtained from supermarkets. Keeping options open on other avenues would permit other vehicles of product distribution to evolve to offer a wider range of product differentiation to broader classes of consumers that have varying attitudes with respect to value-for-money when confronted with the issues of product quality, food safety, and production processes involved.

Largely neglected, correspondingly, has been a systematic investigation of the broader range of differentiated meat and milk products that consumers in domestic markets do demand, for which they may not necessarily turn to the supermarkets to obtain them, and for which they may not necessarily demand formal certification for product quality, food safety, or manner of production. Much of the differentiated products demanded by the wider range of consumers in the domestic market, such as meat or milk products from local varieties of livestock, or eggs from local varieties of poultry, are in fact derived from livestock products that rural smallholders are already producing with the breeds of livestock they have, with the inputs that they use, and with the technology at their disposal. Where there is yet a gap between consumer demands and the supply of such products, these rural smallholders are in a position of comparative advantage to build on what they are already engaged in and exploit that potential, rather than channelling them to fit into input-intensive industrial-type livestock products destined for high-end markets that are screened by extremely tight product and process standards.

Key to matching consumer demand for particular meat and milk products and the supply of the same by smallholders, is an efficient organization of production and distribution, and an effective and recognized institutional mechanism for assuring the quality and safety, and ensuring the identity of such products as claimed to be differentiated. More systematic investigation into these issues should be undertaken. Efficient organization requires predictability and accountability for the sustainability of recurrent transactions. Thus, binding contracts must govern transaction relations. The issue of formality of contracts should not be a prior restriction, but alternative specifications of contracts should be explored to take into consideration the initial smallholder conditions and constraints to create a starting point for the evolution and strengthening of relations that build upon the reliability and trustworthiness of smallholders in delivering their own part of the agreement. These transaction relations may later evolve into formal contracts, but they may also evolve as strengthened informal agreements built on mutually reinforced trust between both parties.

The evolution of alternative institutions that provide guarantee for the identity, quality and safety of differentiated products that are derived from smallholder livestock output should be more systematically explored so as to cover a wider range of differentiated meat and milk products demanded by broader classes of domestic consumers. This would allow the classification of livestock and meat products along a wider spectrum than a simple dichotomy between 'certified' products that have 'passed' some public or private standards versus 'uncertified' products that presumably have 'failed' to pass.

Concerted efforts toward these alternative directions would result in the opening up of more windows for smallholder producers of livestock products to be integrated into the evolving demand patterns for meat and milk products in domestic markets, rather than forcing the issue of making smallholder producers be able to produce the high-end products for export markets and supermarkets whose tight product and process standards are extremely difficult for rural smallholders to hurdle.

Discussion in lecture room

Apply your theoretical knowledge with respect to this paper:
- What is the theoretical proposition of the 'invisible hand' of Adam Smith? Are the assumptions of Adam Smith also true for goods such as clean air or clean water?
- Discuss the relation of formal and informal contracts to economies of scale and social capital. Social capital is a theoretical concept founded by Pierre Bourdieu in the 1960s and later reformulated by James S. Coleman in the 1990s. Do you think that informal contracts are related to the social capital of the involved actors? Conduct desk research on cultural and symbolic capital, the two other forms of capital described by Bourdieu. In which way is consumer behaviour and their purchase behaviour connected to symbolic capital?
- This paper gives an illustrative description about the dynamics of marketing channels. Whole value chains competing against each other are a typical evolutionary step of marketing channels in highly developed markets. 'Vertical Marketing Systems' strive for competitive advantages through their 'size, bargaining power, and elimination of duplicated services' (Kotler and Keller, 2008: 466). Engage in literature research on different forms of vertical marketing systems and discuss their competitive strengths and weaknesses.

References

Baumann, P., 2000. Equity and Efficiency in Contract Farming Schemes: The Experience of Agricultural Tree Crops. Working Paper 139, Overseas Development Institute, October 2000.
Birthal, P.S., Jha, A.K., Tiongco, M., Delgado, C., Narrod, C. and Joshi, P.K., 2006. Equitable Intensification of Market-Oriented Smallholder Dairy Production in India Through Contract Farming. IFPRI Project on Contract Farming of Milk and Poultry in India. FAO AGAL, Rome, Italy.

Catelo, M.A.O. and A. Costales. 2008. Contract Farming and Other Market Institutions as Mechanisms for Integrating Smallholder Livestock Producers in the Growth and Development of the Livestock Sector in Developing Countries. PPLPI Working Paper No. 45, 1 December 2008, FAO, Rome, Italy.

Costales, A., Delgado, C., Catelo, M. A., Lapar, M.L., Tiongco, M., Ehui, S. and Bautista, A.Z., 2007. Scale and Access Issues Affecting Smallholder Hog Producers in and Expanding Peri-Urban Market; Southern Luzon, Philippines. IFPRI Research Report No. 151. IFPRI, Washington, D.C., USA.

Da Silva, C.A. 2005. The Growing Role of Contract Farming in Agri-Food Systems Development: Drivers, Theory and Practice. Agricultural Management, Marketing and Finance Service. FAO, Rome, Italy.

Delgado, C. 1999. Sources of Growth in Smallholder Agriculture in Sub-Saharan Africa: The Role of Vertical Integration of Smallholders With Processors and Marketers of High Value-Added Items 38, Agrekon, pp. 165-189 (Special issue).

Delgado, C., Narrod, C. and Tiongco, M., 2003. Policy, Technical and Environmental Determinants and Implications of the Scaling-up of Livestock Production in Four Fast-Growing Developing Countries: A Synthesis. Final Report of IFPRI-FAO Livestock Industrialization Project: Phase II., International Food Policy Research Institute, Washington, D.C., USA.

Delgado, C.L., Narrod, C.A., Tiongco, M.M., Sant'Ana de Camargo Barros, G., Catelo, M.A., Costales, A.C., Mehta, R., Naranong, V., Poapongsakorn, N., Sharma, V.P. and De Zen, S., 2008. Determinants and Implications of the Growing Scale of Livestock Farms in Four Fast-Growing Developing Countries. IFPRI Research Report. Washington, D.C., USA.

Fairoze, M.N., Achoth, L., Rashmi, P., Tiongco, M., Delgado, C.L., Narrod, C., and Chengappa, P.G., 2006. Equitable Intensification of Market-Oriented Smallholder Poultry Production in India Through Contract Farming. Contract Farming of Milk and Poultry in India: Partnerships to Promote the Environmentally Friendly and Equitable Intensification of Smallholder Market-Oriented Livestock Production. Annex II, Final Report of IFPRI-FAO Contract Farming Project in India, International Food Policy Research Institute, Washington, D.C., USA.

Kirsten, J. and Sartorius K.I., 2002. Linking Agribusiness and Small Farmers in Developing Countries: Is There a New Role for Contract Farming? Development Southern Africa 19: 503-529.

Kotler, P. and Keller, K.L., 2008. Marketing Management. 13th edition, Pearson International Edition, London, UK.

Minot, N., 1986. Contract Farming and its Effect on Small Farmers in Less Developed Countries. Working Paper No. 31. Department of Agricultural Economics, Michigan State University, USA.

Paulson, N.D. and Babcock, B.A., 2007. The Effects of Uncertainty and Contract Structure in Specialty Grain Markets. Selected Paper prepared for presentation at the American Agricultural Economics Association Annual Meeting, Portland, OR, USA, July 29-August 1.

Peterson, H.C. and Wysocki, A., 1997. The Vertical Coordination Continuum and the Determinants of Firm-Level Coordination Strategy. Staff Paper 97-64, Michigan State University, USA.

Pingali, P., Meier, M. and Kwaja, Y., 2005. Commercializing Small Farms: Reducing Transaction Costs. In The Future of Small Farms: Proceedings of a Research Workshop, 25-29 June 2005, Wye, UK. International Food Policy Research Institute. Washington, D.C., USA.

Pingali, P., 2006. Agricultural Growth and Economic Development: A View Through the Globalization Lens. Presidential Address to the 26th International Conference of Agricultural Economists, Gold Coast, Australia, 12-18 August 2006.

Reardon, T. and Timmer, C.P., 2005. Transformation of Markets for Agricultural Output in Developing Countries Since 1950: How Has Thinking Changed? In: R. Evenson, P. Pingali and T.P. Schultz (eds.), Handbook of Agricultural Economics, Vol. 3A. Amsterdam, the Netherlands, pp. 2807-2855.

Rich, K.M. and Narrod C.A., 2005. Perspectives on Supply Chain Management of High Value Agriculture: The Role of Public-Private Partnerships in Promoting Smallholder Access. Draft Working Paper.

Sartorius, K. and Kirsten, J., 2007. A Framework to Facilitate Institutional Arrangements for Smallholder Supply in Developing Countries: An Agribusiness Perspective. Food Policy: 640-655.

Simmons, P., 2002. Overview of Smallholder Contract Farming in Developing Countries. Working Paper 02-04, Agricultural and Development Economics Division of the Food and Agriculture Organization of the United Nations (FAO - ESA). FAO, Rome, Italy.

Son, N.T., Lapar, M.L., Tiongco, M. and Costales, A.C., 2008. Contract Farming for Equitable Market-Oriented Swine Production in Northern Vietnam. A joint ILRI-HAU-IFPRI-FAO PPLPI Project Report.

Tiongco, M., Narrod, C. and Delgado, C. 2006. Equitable Intensification of Market-Oriented Smallholder Dairy and Poultry Production in India through Contract Farming: A Synthesis. Final Report submitted to FAO, June 30, 2006.

Tiongco, M., Catelo, M.A. and Lapar, M.L. 2007. Contract Farming of Swine in Southeast Asia as a Response to Changing Market Demand for Quality and Safety in Pork. A working paper for submission as IFPRI Discussion Paper series.

Promoting neglected and underutilised tuberous plant species in Vietnam

M. Schmidt[1], N.T. Lam[2], M.T. Hoanh[3] and S. Padulosi[4]
[1]*International Dialogue and Conflict Management, Biosafety Working Group, Kaiserstraße 50/6, 1070 Vienna, Austria*
[2]*Hanoi University of Agriculture, Center for Agricultural Research and Ecological Studies, Trau Quy, Gia Lam, Hanoi, Vietnam*
[3]*Plant Resources Center (PRC), Vietnam Agricultural Science Institute (VASI), Vietnam*
[4]*Bioversity International, Via dei Tre Denari 472a, 00057 Maccarese, Rome, Italy*

Abstract

On a global level only about 30 plants provide most of the plant derived energy uptake in human consumption. There are however, about 7,000 plants that have been used for food and agriculture in the history of mankind. Many of these useful species are still used at a local level, but hardly commercialized at large scale due to cultural, agronomic or economic reasons. These so-called neglected and underutilized species (NUS) receive relatively little development and research attention, although they provide large opportunities for society in terms of nutrition security and maintenance of cultural identity. The re-introduction, management and promotion of NUS is highly important to warrant future food security, agricultural diversification and generation of new market opportunities for agricultural products. In this paper we discuss several tuberous NUS for Vietnam, explain why their increased use would be beneficial to improve food security and what could be done to promote these NUS on a local, national and international level.

1. Background

The diversity of cultivated plants plays a major role in peoples' livelihoods by providing food, feed, shelter, medicine and other countless benefits. Crop diversity is a fundamental asset; when there is less diversity available, people, particularly poor people, have fewer options for their nutrition, health and income generation and they are more vulnerable to global events such as new climate patterns. Although there is worldwide a general appreciation, that diversity is important, only a handful of crops dominate the world's agriculture. Of the 7,000 plant species used worldwide in food and agriculture, only 30 crops 'feed the world'. These are the crops that provide 95% of global plant-derived energy-intake (calories) and proteins. Wheat, rice and maize alone provide more than half of the global dietary energy. A further six crops or commodities – sorghum, millet, potato, sweet potato, soybean and sugar (cane/beet) – bring the total to 75% of the global energy intake (FAO, 1996). In fact, 60% of the world's calories are provided by maize, wheat and rice.

When food supplies are analysed at the sub-regional level, however, a greater number of crops emerge as significant. These crops include beans, bananas, lentils, cowpea, yams, groundnut and pea, each contributing to strengthen the diet of millions of the world's poorest (FAO, 1996; Prescott-Allen and Prescott-Allen, 1990). Apart from these crops, many other useful species are also used at local level, but hardly commercialized at large scale due to cultural, agronomic or economic reasons. These so-called neglected and underutilized species (NUS) receive relatively little development and research attention, although they provide large opportunities for society in terms of nutrition security and maintenance of cultural identity (Dawson *et al.*, 2007; Padulosi *et al.*, 2002; Williams and Haq, 2002). These NUS are an important fraction of the world agro-biodiversity portfolio, bearing an undiscovered economic potential and contributing to the reliability of food supply. Conserving and using plant genetic diversity of the world's major crops is certainly vital in meeting the world's future development needs (Figge, 2004); however the re-introduction, management and promotion of neglected and underutilized crop species is also highly important to warrant future food security, agricultural diversification and generation of new market opportunities for agricultural products (Schmidt *et al.*, 2008).

2. The 'long tail' in agriculture

The 4 most important crops alone are cultivated on 50% of the global agricultural area (Figure 1). The 20 most important crops occupy more than 80% of the global agricultural area, leaving hundreds of thousands of o/ther crops – the long tail in agriculture – confined to the remaining 20% (Figure 2). Many hundreds of valuable species still used at a local level, are left at the margins of mainstream agricultural research, marketing and development and, as a result, the potential contribution that these species could make to livelihoods remains largely untapped.

3. Neglected and underutilised species (NUS)

Terms such as 'underutilized', 'neglected', 'orphan', 'minor', 'promising', 'niche' and 'traditional' are often used interchangeably to characterise the range of plant species with under-exploited

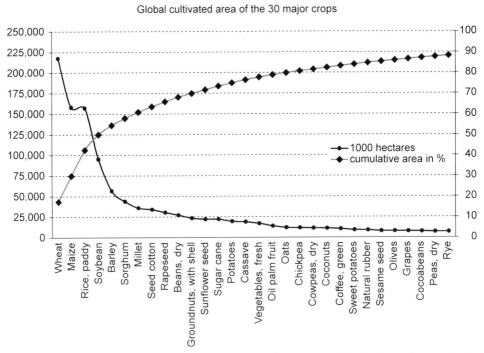

Figure 1. Ranking of the 30 most important crops according to their global cultivated area. The first line (•) shows the total cultivated area per crop (in 1000 hectares), and the second line (♦) shows the cumulative share of total global cultivated area. The four most important crops alone, wheat, maize, rice and soybean, are cultivated on 50% of the global agricultural area. Values are for the year 2007 (FAOSTAT, 2008).

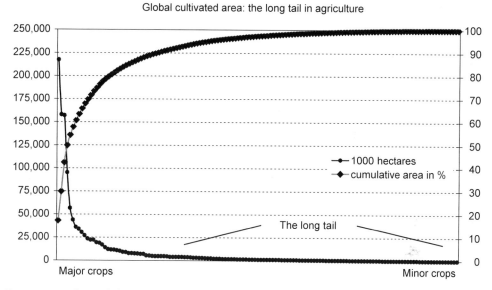

Figure 2. Ranking of the 137 most important crops (categories). The 20 most important crops are cultivated on 82% of the global agricultural area, while the other 117 crops combined contribute to only 18%. Values are for the year 2007 (FAOSTAT, 2008).

potential for contributing to food security, health (nutritional/medicinal), income generation, and environmental services (GFU, 2008). Three of the most relevant institutions working on NUS define them in the following manner (Table 1), stating that these species are under-exploited or overlooked, but yet have potential or play a crucial role for the local communities and the poor.

These neglected and underutilised crop species represent an enormous wealth of agrobiodiversity, have great potential for contributing to improved incomes, food security and nutrition, are strongly linked to the cultural heritage of their places of origin, are mainly local and traditional crops or wild species, tend to be adapted to specific agro-ecological niches and marginal land, have weak or no formal seed supply systems, are collected from the wild or produced in traditional

Table 1. Main definitions of NUS (Bioversity International, 2008; GFU 2008; ICUC, 2008).

Crops for the Future (formerly International Centre for Underutilised Crops, ICUC)	Bioversity International (formerly IPGRI)	Global Facilitation Unit for Underutilized Species (GFU)
We define underutilised crops as plant species that are used traditionally for their food, fibre, fodder, oil or medicinal properties. They have an under-exploited potential to contribute to food security, nutrition, health, income generation and environmental services.	Neglected and under-utilized crop species (NUS) are those that have been overlooked by scientific research and by development workers, and yet play a crucial role in the food security, income generation and food culture of the rural poor.	Underutilized species are those species with under-exploited potential for contributing to food security, health (nutritional/ medicinal), income generation, and environmental services.

production systems, receive little attention from research, policy makers, donors, and consumers, and may be highly nutritious and/or have medicinal properties or other multiple uses (GFU, 2008; Padulosi and Hoeschle-Zeledon, 2004).

4. Agrofolio: benefiting from an improved agricultural portfolio in Asia

Identifying priority NUS in Southeast Asian countries such as China, Cambodia, Thailand and Vietnam was the main aim of the European Commission FP 6 project Agrofolio (Figure 3). The project was carried out in 2006 and 2007 involving nine research organisations from Austria, Spain, Thailand, Vietnam, China and Cambodia (www.agrofolio.eu). The Euro-Asian cooperation intended to support agricultural diversification, income generation and food security on a regional and international level. The project's objective was to elaborate a list of recommendations and strategies for the sustainable use of priority NUS based on a multi-criteria assessment of Asian NUS. The project linked European researchers to their Asian colleagues to investigate the potential sustainable use of NUS, select priority NUS, and to explore new opportunities for niche products.

The identification of promising neglected and underutilised species in China, Cambodia, Northeastern Thailand and Northern Vietnam began with preparing a 'masterlist' containing 260 species, based on farmer interviews and literature review. After an initial pre-selection the team identified 17 NUS for China, 13 for Cambodia, 12 for Thailand, and 22 for Vietnam as of highest priority. These NUS then underwent a multi-criteria and trans-disciplinary assessment involving 511 stakeholders such as scientists, farmers, NGOs and policy makers. The following criteria were used for NUS assessment:
- economic and agronomic competitiveness;
- local and national use and cultural acceptance;
- traditional knowledge;
- scientific knowledge;
- policy & legislation and;

Figure 3. Agrofolio project logo.

- opportunities for national/export niche markets.

Based on the assessment we identified the most promising NUS for each country. The team also identified some limitations for the promotion of so-called priority NUS regarding to some substantial disagreement between the involved stakeholders. In Vietnam, for example, scientists and policy makers had substantially different opinions on NUS compared to farmers. Addressing these ambiguous views will be highly relevant to the development of an improved agricultural portfolio (Schmidt *et al.*, 2008)[10].

5. Priority NUS for Vietnam

Vietnam is considered as one of the countries with the highest agro-biodiversity in the world. Although rice is the dominant crop in Vietnam and consequently, most policies and techniques are focused on rice production. However, environmental issues are arising due to the overuse of fertilizers and pesticides, monocultures and biodiversity degradation, genetic erosion, high risk from yield loss and global climate change. On the other hand, minor crops such as NUS are less concerned by the Vietnamese government policies, although they bring significant contributions to food security and local livelihood, especially for disadvantaged groups or ethnic minority people (Schmidt *et al.*, 2007). Vietnamese agricultural scientists have recognized the potential of some NUS, and are now trying to develop their use on a sustainable basis. In Agrofolio the selection of Vietnamese priority NUS resulted in 22 species, including fruit trees, vegetables and tuberous species (Schmidt *et al.*, 2008). Based on those results Vietnamese agricultural experts started to look closer into a subsection of Vietnamese priority NUS, namely tuberous species.

6. Sustainable development of tuberous NUS in Vietnam

A total of 7 tuberous species have been selected by scientists from the Hanoi Agricultural University and Vietnam Agricultural Science Institute, that can be regarded as priority NUS for Vietnam. Table 2 presents an overview on their status and their exploitable potential.

There are several reasons to propose the above-mentioned tuberous species. They are all traditional species, easy to cultivate, and have good resistance in difficult environmental conditions such as poor nutritional soil, dry or wet land, and reduce erosion. Some species can be intercropped with other species or grow under other trees's canopy such as *Canna edulis*. In addition, these species can be cultivated on fallow land. These species are food crops to be added in daily meals to reduce rice intake, while increasing its amount for export. They can provide important contribution to reinforce human health, especially with regard to glycosuria, liver diseases, etc. They are easy to cultivate in many areas in Vietnam and can contribute to increasing household incomes of local people, in near future, with their potential to be exported.

6.1 Ipomoea batatas *(L.) Lamk (sweet potato)*

For one species, sweet potato, health benefits have especially been investigated (although underutilized, this crop is considered the third most important one in Vietnam after rice and maize). Sweet potato can contribute to prevent cancer (esophagus and colon), and is a beneficial food for diabetics, as it helps to stabilize blood sugar levels and to lower insulin resistance, thus

---

[10] Since the selection of the initial masterlist of 260 species, several additional species were added resulting in a total of 305 NUS that are currently available in the 'Agrofolio database for neglected and underutilized species from China, Cambodia, Thailand and Vietnam' that can be accessed at http://www.agrofolio.eu/db/.

Table 2. Selected tuberous NUS for Vietnam.

Tuberous species (Vietnamese name)	Their status		Exploitable potential	
	Production	Market	Use	Domestic	Export
Ipomoea batatas (Khoai Lang)	Can be cultivated from 2-3 crops per year in some regions in Vietnam	Market niche is narrow (local markets) Tuber can be eaten instead of rice	Leaves used as vegetable Tuber is eaten Feed for livestock	Its leaves are considered as one type of safe vegetable Popular dish is tuber, sometimes it is processed into other types Tubers can replace rice Feed for livestock Health benefits	Tuber in the south Seedling in the north
Dioscorea esculenta (Củ từ)	Can be cultivated 1-2 crops per year by households in some regions in VN	Market niche is narrow	Freshly eaten	Food, sometimes replaces rice, food for export	Food products It can be exported in near future
Colocasia antiquorum (Khoai sọ)	Can be cultivated 1-2 crops per year on dry and marginal land Intercroping with corn, peanut and other vegetable on the field	Market niche is popular/ abundant	Food, sometimes as feed for livestock	Food for people and livestock	If livestock increases, this species could be exported
Colocasia esculenta (Khoai mon)	Can be cultivated 1-2 crops per year on wet land	Market niche is narrow	Leaves and stem is feed for livestock, and tuber is eaten in the rural areas	Feed for livestock	If livestock increases, this species could be exported and raw material for food processing industry such as fibre, baby foods
Maranta arundinacea (Khoai dong)	Can be cultivated one crop per year on dry land (in gardens)	Market niche is narrow	Food (in some areas)	Food processing	It can be processed as special food products Its starch can be exported
Canna edulis (Dong riếng)	Can be planted from February to May	Market niche is narrow	Food, young leaves, stem are feed for cattle	Food processing	It can be processed for special food products
Trapa bispinosa (Củ ấu)	Can be cultivated one crop per year in ponds	Market niche is narrow	Food (in some areas)	Food through marketing	It can be processed for special food products It can be exported

prevent glycosuria. It can also absorb toxin, so it can prevent for inflammation in stomach and bowel. Sweet potato can be regarded as 'safe' vegetables as pesticides or herbicides are rarely applied. Particularly, leaves of sweet potato contain poliphenolistics (Ceffeoylquilic acid, chlorogenic acid, caffeic acid) that have been identified as phenol components. Leaves and tops of sweet potato contain raw protein, Vitamin A, C, mineral substance and fibrous matter. In addition, these also prevent oxidization that causes mutation, have lutein matter to protect eyes and have some substances that regulate physiology (See e.g. Islam, 2006; Kurata *et al.,* 2007; Ludvik *et al.,* 2002; Yoshimoto *et al.,* 2002). Estimations on the agronomic potential indicate that sweet potato cultivation is 50-60 million Vietnamese dong per ha (about 2,300 to 2,760 euro) in Chau Thanh – Tien Giang, Vietnam, which is 2-3 time higher than that of rice cultivation. Efficiency of sweet potato cultivation is higher than that of other tuberous crops within three different species of taro and potato family.

6.2 Dioscorea esculenta *(Lour.) Burk. (yam)*

Yam is planted in many parts of Vietnam as well as in some other Asian countries. It is easy to plant on a variety of soil types. It can be grown under drought conditions and has good resistance towards insect pests. Local people cultivate yam for food, increase income and prevent soil erosion. The yield of yam is on average 10 tons per ha. Yam root is used for food and favourable cooking soup. The root is used for extracting water for vomiting when men were poisoned. It is also used to detoxicate, antidotal, preventing cough and rheumatism.

6.3 Colocasia antiquorum *(L.) Schott (taro)*

Taro is most popular in some provinces in Northern Vietnam, Quang Tri, Soc TRang and Ho Chi Minh city. It can be planted on wet or dry land. Moreover, Taro can grow in drought condition and poor land. Taro is cultivated to protect land and reduce erosion on hill land, and the yield ranges from 8 to 20 ton per ha. Leaves, stem, and roots are used for animal fodder. The root is also used for food and the leaves are prepared as a vegetable. There is also a medicinal use, the fresh root is pounded to cover furuncles, snake bites, and bee stings.

6.4 Colocasia esculenta *(L.) Schott (Indian taro)*

Indian taro is popular in Vietnam, however, it is cultivated mainly in the mountainous regions and midland areas in Northern Vietnam. There are many famous Indian taro varieties in Vietnam such as pink root, red root in Bac Kan province, yellow tuber in Ninh Binh, Yen Bai, Hung Yen, provinces. Farmes can produce about 15 to 20 tons per ha. The root is used for food such as vegetable and processed food for industry, including food for babies. In addition, leaves, stem and roots are used as feed for livestock.

6.5 Trapa bicornis *Roxb. (water caltrops/water chesnut)*

Trapa is grown in ponds or paddy fields, where water is available all year round. Trapa is easy to plant, so local people do not need to spend too much on fertilizers, pesticides and labour. It is planted on a small scale in the Northern Vietnam provinces: Quang Tri, Soc Trang and Dong Thap, and around Ho Chi Minh city. The yield is on average between 15 and 17 tons per ha. There are 7 to 8 harvests per crop. Trapa seed is used as a food or sugar-preserved trapa seeds. Trapa burned (roasted) seeds are used for preventing headache and fever. Cover trapa seeds are used for preventing stomach ulcers.Trapa could become a food product for export to Taiwan.

6.6 Canna edulis *Ker-Gawl. (edible canna)*

Canna's starchy rhizomes are sold to processors in the Vietnamese lowlands to make a type of clear noodle. It is an important crop plant in the mountain regions of Vietnam. It can be cultivated on a variety of soil types and rarely fails to produce yields. Besides, canna can be grown well under other species's canopy, so it is suitable for intercropping with other crops such as peach, plum, and apricot and forest trees. Average yields are about 7 to 20 tons/ha. Canna root is used for food processing, e.g. cooking soup. Young leaves, stem and root residue are good feed for cattle. Canna root also has a medicinal use as tranquillizer and to reduce fever.

6.7 Maranta arundinacea *L. (arrow root)*

Maranta is a popular plant in gardens, fields and hill sides of Vietnam's regions Vinh Phuc, Hoa Binh, Ha Noi. Average productivity is between 10 and 12 tons per ha. The root is rich in starch, which is used for food processing. In addition, Maranta's root is not only food for people and animals but can also be used as a medicinal plant as diuretic.

7. Recommendations to promote tuberous NUS in Vietnam

The above-mentioned tuberous species can be divided into two groups: short-cultivation species and long-cultivation species. *Ipomoea batatas* and *Colocasia esculenta* are in the short-cultivation group that can be cultivated 2, 3, or 4 crops per year depending on usage. The rest are in the long time group that can be cultivated one crop per year. Both groups can be cultivated on dry or wetland, especially *Ipomoea batatas* and *Colocasia esculenta* can be cultivated in the winter season in North Vietnam and other seasons in the Central and Southern provinces.

7.1 Objectives

In order to reduce the amount of rice consumed, products made from tuberous species such as sweet potato and taro should be added as a regular food source in the daily meals. Reducing the share of rice is not only important in order to improve dietary demands, but could well be an economic necessity, as the amount of rice available for export will increase (Vietnam is the second largest rice exporter in the world after Thailand). Moreover, these species can protect soil and reduce erosion in fallow and hill-land. The Government should consider prioritization of selected tuberous species as means to improve food security in Vietnam and elsewhere.

7.2 Proposed solutions

To encourage cultivation of tuberous species, it is necessary to establish adequate marketplaces for starch and tubers with suitable prices. Furthermore the aim should be to create stable markets (demand and production) for these products, and supply chains from producers to stores should be established. The main production areas should be identified to ensure products made from tuberous species are always available for the market. For research and development the following is needed:
- materials for plant breeding;
- increase research nationally and internationally on plant genetic resources of selected tuberous species;
- introduce the list of selected NUS to the list of valuable plants that require conservation and sustainable utilization;

- research ways to improve acceptance of selected priorities of NUS for production and marketing purposes;
- support for local people to exploit, use and conserve NUS in the long run.

7.3 Policy recommendations

On the national Vietnamese level it is recommended that starch from tuberous species should be added to daily food consumption, in order to replace 10-20% of rice ('Do not eat only rice'). This diversification of food sources would not only be beneficial for Vietnamese people in terms of dietary constraints, but the decrease in home rice consumption could be used for export. That way the increase in Vietnamese consumption of selected tuberous NUS would have a beneficial economic effect on rice export; some previous initiatives by the government seem to suit these recommendations. For example the Vietnamese Ministry of Agriculture and Rural Development (MARD) launched the Decree 80/2005/QD-BNN (dated 05 December 2005), which provided the list of valuable NUS for further conservation (1,400 germplasm). Also decree 79/2005/QD-BNN (dated 05 December 2005), issued regulation and term of reference for international exchange of valuable plant species, and several of the priority NUS mentioned here, were included in the Decree 79 (see Schmidt *et al.,* 2007).

It is necessary to create more awareness to encourage consumption from tuberous species as a source of starch. Government and provincial authorities should support initiatives to adjust food proportion that comes from tuberous species to ensure good health, and prevent some diseases.

On the provincial level it is necessary to set up a supply chain, allowing producers to make contracts with channel distributors (marketplace). Also policies to support tuberous plant production and processing should be established, so that these species could be cultivated in suitable areas so that their products can become popular commodities.

Furthermore it is necessary to develop and implement a national strategy for conservation and sustainable use of NUS, in order to ensure the balance between exploitation for basic needs and environmental sustainability. The main focus should be laid on priority NUS, those that have a (potential) high value and/or are at risk of genetic erosion. A classification of NUS based on assessment of potential use and threat assessment (similar to e.g. Schmidt *et al.,* 2008) is needed in order to develop rational policy and regulations for control and conservation of these NUS. Such a national strategy should for example include the following elements:
- Educate and raise public awareness on priority NUS among user groups and especially younger generations.
- Establish and carry out a number of appropriate ex-situ (gene banks) and in-situ conservation programs as well as agronomic, value addition and marketing research on NUS within the whole country.
- Decentralization and empowerment of local authorities and communities to use and promote NUS through an enabling policy environment.
- Promotion of exchange programs with international agencies/organizations with regard to knowledge and marketing aspects related to NUS. The government promotes national and international cooperation programs on assessment and use of native genetic plant resources. For instance, Hanoi Agricultural University has implemented a project named Genetic Resources Policy Initiative (GRPI) with International Institute for Plant genetic Resources in 2005-2006 (Van Dinh *et al.,* 2006).

- Document indigenous knowledge, so as to increase self-esteem and recognition of local communities and culture, safeguarded through the continued use of local crops and their rich associated traditions.

Acknowledgements

The authors were supported by a grant from the European Commission's 6th framework programme 'Diverseeds: networking on conservation and use of plant genetic resources in Europe and Asia', contract 031317. We would like to thank Ms. Pham Thi Dung, from the Center for Agricultural Research and Ecological Studies, Hanoi University of Agriculture, Vietnam, for comments on the paper, and Stefan Glaser for introducing us to the editor-in-chief.

Discussion in lecture room

Apply your theoretical knowledge with respect to this paper:
- Discuss the implications of this paper with consideration to biodiversity. What are the implications with respect to food security for a constantly growing world population? Do you know any local or national initiatives to preserve and collect old plant varieties for food production?
- Read the paper by Costales and Catelo (2010) and use their results to develop policy guidelines and marketing strategies to promote the production and sales of NUS.
- Discuss the importance of the Novel Food Regulation EC 258/97 for the import of NUS into the EU. How is Novel Food defined? Search for the list of applications for authorisation of a specific novel food. Research the case of *Stevia rebaudiana* (available at: http://ec.europa.eu/food/food/biotechnology/novelfood/app_list_en.pdf).

References

Bioversity International, 2008. Neglected and Underutilized Species –Overview. Available at: http://www.bioversityinternational.org/scientific_information/themes/neglected_and_underutilized_species/overview/. Accessed January 2009.

Costales, A. and Catelo, M.A.O., 2010. Market access for smallholder livestock producers under globalised food trade: An analysis of contract farming and other market-linking institutions in India, Thailand, Vietnam and the Philippines. In: R. Haas, M., Canavari, B. Slee, C. Tong and B. Anurugsa (eds.), Looking east looking west: organic and quality food marketing in Asia and Europe. Wageningen Academic Publishers, Wageningen, the Netherlands, pp. 173-181.

Dawson, I.K., Guarino, L. and Jaenicke, H., 2007 Underutilised Plant Species: Impacts of Promotion on Biodiversity. Position Paper No. 2. International Centre for Underutilised Crops, Colombo, Sri Lanka.

FAO, 1996: Report on the State of the World's Plant Genetic Resources for Food and Agriculture, prepared for the International Technical Conference on Plant Genetic Resources, Leipzig, Germany, 17-23 June 1996. Food and Agriculture Organization of the United Nations, Rome, Italy.

FAOSTAT, 2008. Production Statistics of Crops. Available at: http://faostat.fao.org/site/567/default.aspx#ancor. Accessed January 2009.

Figge, F., 2004. Bio-folio: Applying Portfolio Theory to Biodiversity. Journal of Biodiversity and Conservation 13: 827-849.

GFU, 2008. When we say 'underutilized species' this is what we mean. Available at: http://www.underutilized-species.org/default. Accessed January 2009.

ICUC, 2008. What are underutilised crops? Available at: http://www.icuc-iwmi.org/. Accessed January 2009.

Islam, S., 2006. Sweetpotato (Ipomoea batatas L.) Leaf: Its Potential Effect on Human Health and Nutrition. Journal of Food Science 71: R13 - R121.

Ludvik, B.H., Mahdjoobian, K. and Waldhaeusl, W., 2002. The effect of Ipomoea batatas (Caiapo) on glucose metabolism and serum cholesterol in patients with type 2 diabetes: a randomized study. Diabetes Care 25: 239-240.

Kurata, R., Adachi, M., Yamakawa, O. and Yoshimoto, M., 2007. Growth suppression of human cancer cells by polyphenolics from sweetpotato (Ipomoea batatas L.) leaves. Journal of Agricultural and Food Chemistry 10: 185-90.

Padulosi, S., Hodgkin, T., Williams, J.T. and Haq, N., 2002. Underutilized crops: trends, challenges and opportunities in the 21st Century. In: J.M.M. Engels, V. Ramanatha Rao, A.H.D. Brown and M.T. Jackson. Managing Plant Genetic Diversity. Rome, Italy IPGRI (International Plant Genetic Resources Institute). 323-338.

Padulosi, S. and Hoeschle-Zeledon, I., 2004. Underutilized plant species: what are they? LEISA 20: 5-6.

Prescott, A. and Prescott, A., 1990. How many plants feed the world? Conservation Biology 4: 365-374.

Schmidt, M., Banterng, P., Chuong, S., Dung, P.T., Glaser, S., Hager, V., deKorte, E., Lam, N., Li, Y.-H., Meyer, A., Phuong, N.T., Polthanee, A., Qiu, L.Y., del Rio Murillo, A., Sánchez, V., Sophoanrinth, R., Wei, W., Zhang, Z.Y. and Zhou, H.F., 2007. AGROFOLIO – Benefiting from an Improved Agricultural Portfolio in Asia. Final Report. FP6-2002-INCO-DEV/SSA-1–026293. Available at: http://www.agrofolio.eu/index.php?page=pub-final-report. Accessed January 2009.

Schmidt, M., Wei, W., Polthanee, A., Thanh Lam, L., Chuong, S., Qiu, L.-J., Banterng, P., Thi Dung, P., Glaser, S., Gretzmacher, R., Hager, V., de Korte, E., Li, Y.-H., The Phuong, N., Ro, S., Zhang, Z.-Y. and Zhou, H.-F., 2008. Ambiguity in a trans-disciplinary stakeholder assessment of neglected and underutilized species in China, Cambodia, Thailand and Vietnam. Biodiversity and Conservation 17: 1645-1666.

Van Dinh, N., Fadda, C., Ngoc Kinh, N., Quang Hung, H., Halewood, M., Lettington, R. and Thi Ngoc Hue, N. (eds.), 2006. Proceeding of the workshop genetic resources policy initiative I. Agricultural Publishing House, Hanoi, Vietnam.

Williams, J.T. and Haq, N., 2002. Global research on underutilized crops. An assessment of current activities and proposals for enhanced cooperation. ICUC, Southampton, UK.

Yoshimoto, M., Yahara, S., Okuno, S., Islam, M.S., Ishiguro, K. and Yamakawa, O., 2002. Antimutagenicity of mono-, di-, and tricaffeoylquinic acid derivatives isolated from sweetpotato (*Ipomoea batatas* L.) leaf. Bioscience Biotechnology and Biochemistry 66: 2336-41.

Consumer perceptions of organic foods in Bangkok, Thailand[11]

B. Roitner-Schobesberger[1], I. Darnhofer[2], S. Somsook[3] and C.R. Vogl[4]

[1]University of Natural Resources and Applied Life Sciences, Department of Applied Plant Sciences and Plant Biotechnology, Vienna, Gregor Mendel Strasse 33, 1180 Vienna, Austria

[2]University of Natural Resources and Applied Life Sciences, Department of Economic and Social Sciences, Vienna, Feistmantelstrasse 4, 1180 Vienna, Austria

[3]Thammasat University, Pathumthani, Department of Agricultural Technology, 12121 Thailand

[4]University of Natural Resources and Applied Life Sciences, Department of Sustainable Agricultural Systems, Vienna, Gregor Mendel Strasse 33, 1180 Vienna, Austria

Abstract

In response to food scares related to high levels of pesticide residues sometimes found on vegetables and fruits, consumers in Thailand increasingly demand 'safe' foods. This has resulted in a number of initiatives and labels indicating 'pesticide safe' vegetables. However, the pesticide-residue problem has proved enduring. This opens a market opportunity for organic foods, which are produced entirely without using synthetic chemicals. As little is known on consumer perception of organic foods in Thailand, a survey was conducted in Bangkok. More than a third of the 848 respondents reported having purchased organic vegetables or fruits in the past. The main reasons for purchasing organic products are that consumers expect them to be healthier, that organic products are environmentally friendly. The respondents who have bought organic vegetables tend to be older, have a higher education level and a higher family income than those who have not bought them. The main barrier to increasing the market share of organic vegetables is that consumers do not clearly differentiate between the various 'pesticide safe' labels and the organic labels. Informing consumers about unique characteristics of organic production methods, the strict inspection and required third-party certification might be a promising strategy to develop the market for organic vegetables in Thailand's urban centers.

1. Introduction

Interest in organically produced food is increasing throughout the world in response to concerns about intensive agricultural practices and their potential effect on human health as well as on the environment. In Thailand, as in many Asian countries, the rapid socio-economic development was accompanied by a modernization and industrialization of the agri-food production. The Thai government has promoted an industrial, export-oriented agriculture, characterized by a heavy reliance on synthetic chemicals to protect crops against weeds, pests and diseases and thus leading to improved productivity (UNDP, 2007). However, insufficient farmer training has lead to the inadequate use of pesticides, i.e. the recommended application levels and application frequency are not always followed, nor is the pre-harvest interval strictly observed (Chunyanuwat, 2005). Also, farmers use synthetic chemicals that are classified as 'extremely hazardous'[12] and even pesticides that are banned in Thailand (IPM-DANIDA, 2004; Posri et al., 2007). This has lead to

---

[11] Reprinted from B. Roitner-Schobesberger, I. Darnhofer, S. Somsook, C.R. Vogl, Consumer perceptions of organic foods in Bangkok, Food Policy 33: 112-121, Copyright (2008), with permission from Elsevier.

[12] The World Health Organization (WHO) classifies pesticides by hazard based on their lethal dose. Each pesticide is put into one of four classes, from Ia 'extremely hazardous' (i.e. very toxic) to III 'slightly hazardous' (i.e. caution) (WHO, 2006). Because farmers in developing countries often do not have the training or the equipment to handle pesticides safely, FAO recommends that pesticides classified as 'extremely hazardous' should not be used in developing countries (Eddleston et al., 2002).

cases of pesticide poisoning as well as levels of pesticide residues on foods which are above the allowed maximum limits (Chunyanuwat, 2005).

As a response, a range of measures have been taken by the Thai Government as well as NGOs (Posri *et al.*, 2007). To reduce the pesticide use various on-farm 'safe food' projects have been initiated, e.g. by Royal Projects or by the Department of Agriculture. Products originating from these projects carry labels indicating that residues are below the maximum allowable level (Table 1). However, despite increased controls, tests carried out by the Department of Agriculture as well as by the Ministry of Public Health showed that maximum allowable residue levels were still exceeded (IPM-DANIDA, 2003). Even some vegetables labelled as 'pesticide free' were found to be contaminated with pesticide residues above maximum allowable limits (Hardeweg and Waibel, 2002). The resulting risk from pesticide residues in vegetables is well known to consumers through recurring coverage in mass-media and through discussions in public forums organized by the Ministry of Public Health (Kramol *et al.*, 2006). The uncertainty over the magnitude of the risks associated with the contamination of food was exacerbated by other food crises, especially avian flu (Tiensin *et al.*, 2005) and residues of nitrofuran, a banned antibiotic, in shrimp and poultry (Foodmarket, 2003). These food scares have unsettled consumer confidence in Thailand's food industry and in the government's food regulatory agencies (Delforge, 2004).

With increasing awareness of the severity of the domestic problem as well as increasing pressure by international trading partners to comply with international standards, the Thai government overhauled its approach to food safety. The Cabinet passed a 'Road map of food safety', which serves as a framework for the control of food and agricultural products throughout the food chain. To communicate its new approach to consumers, the Cabinet declared the year 2004 'Thailand's food safety year' (Srithamma *et al.*, 2005).

Within this social and political environment, the demand for (really) safe foods is likely to increase. Indeed, highly publicized food safety incidents can lead to lasting changes in food purchasing behavior (Buzby, 2001). Since organic products are produced without the use of synthetic pesticides, they are well placed to answer the consumers' food safety concerns. Indeed, analyses of pesticide residues in produce in the US and Europe have shown that organic products tend to have significantly lower residues than conventional products (Lotter, 2003; Magkos *et al.*, 2006).

Although there is a large number of studies on organic consumers in Europe and other western countries, notably the USA, Australia and New Zealand (see, Lohr, 2001; Thompson, 1998; Yiridoe *et al.*, 2005; Zanoli, 2004 for reviews), little is known on consumer's perception of organic foods in Asia (ACNielsen, 2005; Moen, 1997; Nelson, 1991; Zhang, 2005). This study aims to address this gap by providing first insights into the knowledge about organic foods and the reasons consumers have to purchase or not to purchase organic vegetables in Bangkok. The focus is on fresh vegetables (which include fresh herbs) because it is the most widely available organic product group, and because consumers are often most concerned about residues on fresh vegetables (Zhang, 2005), since they play a key role in Asian and Thai cuisine (Kosulwat, 2002; Veeck and Veeck, 2000).

2. Research method

Together with Thai experts on organic farming, a questionnaire was designed to gather exploratory data on consumer perception of organic foods (see Roitner-Schobesberger, 2006). The questionnaire was pre-tested in English, then translated into Thai and pre-tested again.

Table 1. Major safe food labels in Thailand (Eischen et al., 2006; IPM-DANIDA, 2003; Srithamma et al., 2005).

Label	Title of the label (including the translation of the Thai text on the label)	Origin and description
	Hygienic food (Pilot project for hygienic fresh vegetables and fruits – hygienic fresh vegetables and fruits – Department of Agriculture)	The label was originally used on produce originating from the 'Hygienic fresh fruit and vegetable production pilot project' that was initiated in 1991 by the Dept. of Agriculture (Ministry of Agriculture and Cooperatives). In the project, the use of synthetic chemicals is regulated and controlled. The label is meant to be replaced by the new 'Food quality and safety' label (below).
	Food quality and safety (Ministry of Agriculture and Cooperatives – 'Safe Food')	This quality and safety certification label is given to agricultural commodities and food products that conform to the standards established by the National Bureau of Agricultural Commodity and Food Standards (Ministry of Agriculture and Cooperatives).
	Pesticide-safe vegetables (Quality certification for toxic substances control – Department of Medical Science – Ministry of Public Health)	The Ministry of Public Health assigns the label to retailers who conduct tests for toxic substances before selling the products. The label is used on fresh food products that meet the safety requirements of the Ministry of Public Health.
	Organic Thailand (Organic Products)	The official organic label by the Department of Agriculture. It indicates that the product has been produced according to the organic farming standards set by the Department of Agriculture.
	Organic Agriculture Certification Thailand (ACT)	These products are certified organic by Organic Agriculture Certification Thailand (ACT), a private certification body accredited with IFOAM since 2001.
	IFOAM	Label of the International Federation of Organic Agriculture Movements (IFOAM). Although IFOAM does not certify organic farms itself, the label can be used by certifying bodies accredited by IFOAM.

Before finalizing the questionnaire, the Thai version was translated back into English to ensure that the questions had retained their original meaning. The questionnaire was divided into three parts: first the respondent's general knowledge of 'safe food' labels commonly found on vegetable products (Table 1) are explored. In the second part of the questionnaire, respondents who have indicated that they had heard the term 'organic' were presented with statements regarding organic agriculture and food, and asked whether they agree with these statements. Also, reasons to purchase or not to purchase organic products were assessed. In the last part of the questionnaire basic demographic data was collected.

The data was collected in late April and early May 2005. Five supermarkets and two health food stores carrying organic foods as well as a range of fruits and vegetables displaying the major 'safe food' labels were selected for the interviews. Shops both at the centre and at the outskirts of Bangkok were selected to ensure that a range of customer types are included in the study. Twelve Thai students were trained to administer the questionnaire personally to 848 customers. The sample is mostly a convenience sample, i.e. customers were approached randomly. However, to reduce a potential non-coverage bias, a sampling frame covering age, gender and education level was used. Although there is no way of knowing if those included are representative of the overall population, the survey is still expected to give a first overview of relevant issues and to allow to derive insights into the perception of organic foods by consumers in Bangkok.

The collected data was summarized using descriptive statistics. To analyze differences between consumer types, the respondents were divided into three groups: those who had never heard of 'organic' (the 'never heard organic': 275 respondents, i.e. 33%), those who have heard of organic, but never purchased any organic products (the 'organic non-buyers': 224 respondents, i.e. 27%) and those who have heard of organic and have purchased organic foods in the past (the 'organic buyers': 333 respondents, i.e. 39%)[13]. The significance of differences between the three consumer groups was established using contingency tables and the Chi-Square test (at a 5%-level of significance). The issues mainly motivating consumers to purchase organic food were identified with an exploratory factor analysis (principal component analysis, varimax rotation).

3. Results of the consumer survey

In most demographic variables, there is a significant difference between the three groups (Table 2). Consumers with a lower income and a lower level of education are least likely to have heard of organic agriculture. Conversely, those who have a higher income and hold an academic degree are more likely to have bought organic products in the past. Indeed, most 'organic buyers' (58%) tend to have an academic degree, whereas 46% of 'non-buyers' and 33% of 'never heard' hold a Bachelor degree or higher academic degree. Regarding income, 58% of 'organic buyers' have a monthly family income of over 30,000 Thai bath, compared to 45% of the 'non-buyers' and 38% of the 'never heard'. This shows that there is a strong relationship between education level and income.

'Organic buyers' tend to be older than the other two groups and likely to be men. Of the organic buyers, 25% are over 40 years old and only 22% are under 30 years old, whereas in the two other groups, more than 40% of the respondents are under 30 years old. 'Organic buyers' are slightly more likely to be men: 41% of male respondents said that they had purchased organic products in the past, compared to 40% of the women. There is no statistical difference in the household income level between the interviewed men and women. However, there is a statistical difference

---

[13] For the missing 1% the answer to the question: 'Have you ever purchased organic products?' was not recorded.

Table 2. Demographic characteristics of respondents.

Variable	Number of respondents[1]	Never- heard organic	Organic non-buyers	Organic buyers	Significance of the difference between the groups[2]
Average age (in years)	845	34.3	34.5	42.0	*
Sex					*
Female	551	37.0%	23.4%	39.6%	
Male	280	25.0%	33.9%	41.1%	
Children in the household?					ns
Yes	409	29.8%	26.7%	43.5%	
No	419	36.0%	27.2%	36.8%	
Highest education level					*
Secondary school	174	51.1%	25.9%	23.0%	
High school	265	34.7%	28.3%	37.0%	
BSc	291	23.0%	28.5%	48.5%	
MSc or more	98	25.5%	21.4%	53.1%	
Family income per month[3]					*
< 10,000 bath	91	51.6%	30.8%	17.6%	
10,000-20,000 bath	217	43.8%	25.3%	30.9%	
20,001-30,000 bath	128	28.9%	30.5%	40.6%	
> 30,000 bath	380	24.0%	26.3%	49.7%	

[1] Some information was missing on some questionnaires, thus not all categories add up to 848 respondents.
[2] Significance: * = α≤0.5; ns: not significant.
[3] The National Office for Statistics in Thailand reports that the average monthly family income in Bangkok in 2004 was approx. 29,800 bath (100 bath ≈ 2.07 euro).

in the education level: whereas approx. 35% of men and women have Bachelor degree, 16% of men hold a Masters degree or higher compared to only 10% of women. Also, the men are slightly more likely to have a child living in their household (52% of interviewed men, compared to 48% of women) and to be older (men were on average 40 years old, women 37 years old).

Although the 'organic buyers' are more likely to have children living in their household, the relationship is not significant. Of the organic buyers 54% reported having a child living in their household, compared to 49% of non-buyers and 45% of 'never heard'. The age of the youngest child living in the household is similar for all respondents (6.7 years for 'organic buyers', 6.3 years for 'non-buyers' and 5.3 years for 'never-heard').

Asked about their concerns regarding pesticide residues on vegetables and fruits, over half of the respondents stated that they were 'very much' concerned (Table 3). Not surprisingly there is a significant relationship between the consumers who are concerned about pesticide residues in food products and those purchasing organic foods: of those who are 'very much' concerned, 50% are 'organic buyers', whereas 26% belong to the group 'never heard organic'.

The respondents were also asked whether they are concerned about the use of GMOs (Table 3). Only some 10% stated that they were 'very much' concerned, which might be linked to the fact

Table 3. Concerns about pesticide residues and GMO (genetically modified organisms) in percentages (N=848).

Question	Very much	Often	Sometimes	Not at all	No answer
Are you concerned about pesticide residues on vegetable and fruits?	52.0	18.2	19.1	10.6	0.1
Are you concerned about the use of GMO in food products?	10.0	9.4	27.5	46.4	6.7

that seeds of genetically modified crops are banned in Thailand and thus there has been only limited public debate about GMOs in the media.

The respondents were presented with six labels commonly found on vegetables (Table 1). The label that was recognized by more than half of the respondents was the 'hygienic food' label (Table 4). All other labels were not well known, with only about 10% recognizing the two organic labels. Of those respondents who stated that they know one of the six labels, more than half were organic buyers (Table 4). It thus seems that organic buyers are more aware of food labels. When asked to state what they think the label stands for, respondents essentially repeated the words written on the label with little or no additional information. Thus, even if the respondents state that they 'know' the label, it does not mean that they have detailed information on the background of the label, e.g. the criteria which must be fulfilled for the food to be awarded the label, or the organization responsible for overseeing the label. This might explain why 15% of respondents said that they 'know' the label 'Organic ACT', (Agriculture and Organic Certification Thailand) although they have never heard the term 'organic'.

Those respondents who recognized the labels and stated that they knew about the meaning of the label were also asked whether these labels indicated organic production methods. Some 80% of respondents designated the two organic labels, containing the word 'organic' on the label, as organic. More than half also thought that the 'hygienic food' label and the 'pesticide-safe' label indicated produce from organic agriculture.

Table 4. Distribution of respondents by knowledge of labels (N=848).

Label	Number of respondents knowing the label	Percentage of respondents knowing the label by group		
		Never heard organic	Organic non-buyers	Organic buyers
Hygienic food	444	23.4	27.0	49.6
Pesticide-safe	98	20.4	26.5	53.1
Food quality and safety	97	15.5	30.9	53.6
Organic Thailand	87	11.5	19.5	69.0
Organic (ACT)[1]	83	15.7	20.5	63.8
IFOAM[2]	21	14.3	33.3	52.4

[1] Agriculture and Organic Certification Thailand
[2] International Federation of Organic Agriculture Movements

This lack of discernment might be related to the fact that although most supermarkets have distinct sections for labelled and non-labelled vegetables, they tend not to distinguish further between 'hygienic', 'safe' and 'organic' products. Even health food shops usually carry organic products together with labelled conventional items, with little information on the different production systems (Panyakul, 2002).

The respondents themselves are aware of their limited knowledge regarding organic agriculture: 52% of those who have heard the term 'organic' said they were not sure what it meant. Even those who purchase organic products do not feel well informed: 40% of 'organic buyers' said they only know 'a little' about the meaning of organic, 5% said they know 'a lot'. Nonetheless, the 'organic buyers' feel better informed than the 'organic non-buyers': of the 257 respondents who said that they know 'a little' about the meaning of organic, 71% were 'organic buyers' and 29% 'organic non-buyers'. Of the 17 respondents who said they know 'a lot' about organic, 88% were 'organic buyers'.

The respondents, who had heard of organic farming, were presented with 13 statements and asked whether they though these statements were true, false or whether they did not know if the statement is true or false (Table 5). The survey shows that respondents are convinced that organic farming is good for the environment and that organic foods are healthy. However, the responses also show that the consumers are unsure about the differences between the agricultural production methods underlying 'hygienic' or 'safe' labels and those of organic farming. Although 72% of respondents agree with the statement 'organic farming does not use synthetic pesticides or herbicides', 54% agree with the statement 'organic farming uses synthetic pesticides, but less than

Table 5. Assessment of statements about organic farming by the respondents who have heard of 'organic' in percentages (N=549).

Statement	Yes, I agree	No, I do not think so	I don't know
Organic farming is good for the environment	90.0	1.3	6.0
Organic products are healthy	88.0	3.5	5.5
Organic products do not carry pesticide residues	72.0	12.9	12.6
Organic farming does not use synthetic pesticides or herbicides	72.0	11.7	13.5
Organic products are produced without using chemical fertilisers	70.1	11.8	14.8
Organic farming is the same as natural/traditional farming	65.2	16.8	13.7
You can trust a product that carries an organic label and/or organic certificate	64.9	17.9	13.8
The rules for organic production are stricter than for other production methods (e.g. safe)	57.4	15.9	23.3
Organic farming uses synthetic pesticides, but less than other production methods	53.7	28.6	14.9
Organic food products never contain GMOs	51.2	12.9	32.4
The production and processing of organic products is strictly controlled	47.5	22.8	26.4
There is no difference between organic products and hygienic products	43.0	31.7	21.7
Organic is just a marketing gag/promotion	42.4	41.9	12.4

Note: the missing % did not give an answer.

other production methods'. Asked directly about the difference between organic and hygienic products, 43% said that there is 'no difference'. Still, the respondents are somewhat aware that organic farming is based on defined standards and that farms need to be certified. Indeed, 47% agree with the statement 'the production and processing of organic products is strictly controlled' and 42% disagree with the statement that 'organic is just a marketing gag'.

Those respondents who had purchased organic products in the past (333 respondents) were asked about their motives. The most important motive is the expected positive health effects (a reason for 93% of the 'organic buyers'). These expected positive health effects may be related to the absence of pesticide residues, as 92% of the 'organic buyers' said that they purchase organic products because 'they do not contain pesticides/have lower residues'. Some 84% of 'organic buyers' purchase organic products because they are 'good for the environment', which reflects the high level of agreement with organic farming being environmentally friendly by those respondents who had heard of 'organic' (Table 5). Further reasons to purchase organic products were because 'they are fresher than the other products' (54%), because 'they have a better taste' (29%) and because 'I just wanted to try them/try something new' (22%).

The reasons for purchasing organic products were subjected to a Principal Component Analysis (PCA). It yielded three principal components with an eigenvalue greater than 1. Together they explain 60% of the variance (Table 6). The three principal components indicate that there may be three main reasons for purchasing organic food: out of health concerns, out of curiosity or because they are tastier than other foods. The degree of cohesion within each component showed that the first principal component has an acceptable alpha (0.7) and thus displays a good internal consistency. The second principal component has a low alpha (0.4) which may be due to the fact that it encompasses only three items.

The respondents who had purchased organic products were also asked about the availability of organic products and how often they purchase organic products. The majority (58%) of 'organic buyers' were satisfied with the range of organic products available at supermarkets. However an important share (41%) said they would like to buy more organic products, especially a wider range of vegetables and fruits. Regarding their purchasing habits, 51% of the 'organic buyers' stated that they purchase organic products weekly, 24% said once per month and 23% said they purchase organic foods less than monthly.

Table 6. Principal component analysis of the reasons for organic food purchase.

Principal component name	% explained variance	Cronbach's α	Items	Loading
Healthy & environmentally friendly	29.41	0.711	Organic food is good for my health	0.822
			Organic food is good for my children	0.794
			Organic food does not contain pesticides	0.696
			Organic food is good for the environment	0.612
Fun & fresh	18.68	0.448	I just wanted to try them/try something new	0.795
			It is trendy to buy organic products	0.671
			Organic food is fresher than the other products	0.558
Tasty	12.51	--	Organic products have a better taste	0.955

The 'organic buyers' were asked how they rate the prices of organic products. In April 2005 the price difference between organic and non-labelled conventional vegetables in Bangkok varied between 50% (for white cabbage) and 400% (for small hot chilies), with most organic vegetables (e.g. broccoli, eggplant, beans, kale, cucumber) having a price premium of 100-170% above conventional products. Compared to hygienic vegetables, organic produce is approx. 50% more expensive (Chaivimol, 2003). Despite the price difference, nearly 60% of the 'organic buyers' said that the price of organic products was not a problem.

The 'organic non-buyers', i.e. those respondents who had not previously bought organic products (208 respondents, i.e. 39% of those who had heard of 'organic') were asked why they do not purchase organic products. The main reason was that they 'don't know what organic means' (51%). Further reasons were that 'hygienic/safe is enough for me' (39%); 'they are too difficult to get' (39%); that they 'don't think there is anything special about them which justifies a higher price' (30%) and that they are 'too expensive' (29%). Some (14%) stated that they 'do not trust the label/do not think it is really organic'.

4. Discussion

The consumers of organic vegetables in Bangkok tend to be older, hold an academic degree and have a higher income than those not purchasing organic products. This profile is similar to results from studies in Western countries (e.g. Lockie *et al.*, 2002; Padel and Foster, 2005; Thomson, 1998). A study in Northern Thailand also found that willingness to pay for 'safe' vegetables increases with age and income (Posri *et al.*, 2007). Contrary to reports in the literature (e.g. Lockie *et al.*, 2002; McEachern and McClean, 2002; Thomson, 1998), men in Bangkok seem to be more likely to purchase organic foods than women. This might be due to men being willing to pay a higher price premium for organic products than women, as suggested by Wandel and Bugge (1997). Also, the effect of gender is likely to be interlinked with education level, since the men included in the survey were more likely to hold an academic degree.

There are three main motives to purchase organic food in Bangkok: the expected health benefits, the attraction of new and fashionable products and the search for tastier products. Health benefits have been reported as a main motive for purchasing organic food by most studies (Lockie *et al.*, 2002; Magnusson *et al.*, 2003; Padel and Foster, 2005; Yiridoe *et al.*, 2005; Zanoli, 2004). As Lockie *et al.* (2002) point out, health is the one aspect consumers are least willing to compromise. In Thailand, as this study confirms, the health aspect is closely associated with the residues from synthetic chemicals used in agriculture. Indeed, organic products generally have a lower level of pesticide residues (Baker *et al.*, 2002). The principal component analysis reinforces both the importance of health as a motive and the link between health concerns and the fear of pesticide residues. The second important motive to purchase organic food is the consumer's search for new, trendy and fresh products. Freshness is generally a key criterion in the purchase of vegetables and fruits (Péneau *et al.*, 2006; Sakagami *et al.*, 2006) and if Thai organic vegetables can score highly on this criterion, they are more likely to be purchased. The third important motive is related to the better taste of organic vegetables. This is in line with other studies: many organic buyers believe that organic produce tastes better than conventionally grown produce, even if sensory evaluations have yielded inconsistent results (Fillion and Arazi, 2002; McEachern and McClean, 2002; Zhao *et al.*, 2007).

The share of respondents who report having purchased organic products in the past (39%) is close to the 33% of respondents in Asia-Pacific who stated that they 'regularly' purchase organic vegetables (ACNielsen, 2005). The market for organic foods is thus potentially large. However, to be able to tap this potential customer base, it may be necessary to clearly position organic foods as distinct from other 'safe food' labels. Indeed, the study shows that organic label does not have a clear profile for consumers.

The main barrier to purchase organic products is the lack of information consumers have on organic farming methods. Indeed, the survey shows that the main reason for not buying organic products was that they 'do not know what organic means'. Two factors might help explain this lack of knowledge: the lack of information on agricultural production methods and the claims by competing 'safe food' labels that seem to promise the same health benefits expected from organic labels.

Basic information on agricultural production methods, e.g. the amount and type of fertilizers and pesticides used as well as intensity of quality control during and after production is needed to understand the specificity of organic production. However, most urban consumers might not have this detailed knowledge of agricultural practices (Yiridoe *et al.*, 2005). Indeed, with the sources of food being ever more remote, urban consumers tend to have a very partial picture of growing conditions and production practices, which Jaffe and Gertler (2006) have called 'consumer deskilling'. This makes it difficult for farmers who choose production practices that are safer for consumers, to educate their potential customers (Jaffe and Gertler, 2006). In the case of Thailand, educating consumers is made all the more difficult by the wide variety of competing 'safe food' labels. The terms used on these labels include 'hygienic food', 'safety food', 'quality food', 'non-toxic food', 'health food', 'chemical-free', 'pesticide-free' and 'hydroponic' (see Eischen *et al.*, 2006; Kramol *et al.*, 2006; Posri *et al.*, 2007). For consumers the claims made by these labels are very similar to the health benefits they expect from organic products. It is thus not clear what additional benefits organic products have, leading 43% of respondents to state that there is 'no difference between organic and hygienic produce'.

This lack of awareness of the standards covered by organic farming is not specific to Thailand (e.g. Hoogland *et al.*, 2007) and the organic labels are barely recognized by the respondents. However, awareness of the organic label can increase the probability that a consumer would be willing to pay a premium for organic foods (Batte *et al.*, 2007). This does not mean that knowledge about organic farming methods and awareness of the organic labels necessarily translates into purchase. However, consumers who have never heard of 'organic' or who are not aware which labels truly indicate organic products are not likely to purchase them. It thus seems that in Thailand as elsewhere (see Yiridoe *et al.*, 2005; Zanoli, 2004), the lack of consumer knowledge about the specificities of organic agriculture, and thus the criteria covered by the organic label, can be a key issue hampering the development of the demand for organic foods.

Another factor often mentioned as limiting the market share of organic products is price, especially the price difference between organic and conventional products (Lohr, 2001; Padel and Foster, 2005; Zanoli, 2004). The current premium for organic products in Thailand is approximately 50% above the price of foods with a 'safe' label. This is higher than the premium of 10-20% over conventional products reported as acceptable by studies on the willingness-to-pay for organic products (see review by Yiridoe *et al.*, 2005). However, in Thailand as elsewhere, it is unclear to what extent price is really a key factor in the choice between organic and conventional products. The study has shown that of the 'organic buyers', 60% do not see price as limiting factor, and only 29% of the 'organic non-buyers' mention it as a reason for not purchasing organic products. Indeed, organic products that have a premium higher than 20% over comparable conventional products get purchased in many countries (Hamm and Gronefeld, 2004, Yiridoe *et al.*, 2005). Also, there is substantial research showing that consumers routinely buy a wide range of products without knowing their price (Grunert, 2005). Thus, the price of organic vegetables in Bangkok is not likely to be a key issue limiting sales.

5. Outlook

The survey of potential and actual organic consumers in Bangkok allows us to derive some recommendations both at the policy level, i.e. measures to support the development of organic farming, and at industry level, i.e. strategies to increase the market share of organic foods. At the government policy level it would seem advisable to reduce the confusing multitude of 'safe food' labels. The confusing abundance of 'safe food' labels is mostly the result of uncoordinated policies and initiatives by various Ministries and government agencies (Kramol *et al.*, 2006; Posri *et al.*, 2007). It would seem useful that the Department of Agriculture, the Department of Agricultural Extension and the Ministry of Health agree on one label indicating that foods comply with the maximum allowable residue limits. This label should be based on a common control and inspection method. The label should be available to all farmers complying with the standards, not only to those participating in pilot projects. A first step in this direction was the establishment of the 'Food quality and safety' label. However, although this label was introduced on the market, the others (e.g. 'hygienic food' or 'pesticide-safe vegetable) were not simultaneously removed from the market so that this new label has added to the confusion, not reduced it.

Reducing the number of 'safe food' labels would also allow organic labels to be clearly positioned as indicating more restrictive standards. Organic could then be communicated as indicating that due to the production methods the residues are much lower than the maximum allowable residue limits, and as offering additional benefits, particularly regarding the environmental impact. On the domestic market in Thailand there are currently two major organic labels which are related to two different standards and two certifiers (the Department of Agriculture and Organic Certification Thailand (ACT)). It is debatable whether the way forward lies in only one national organic label (as is implemented in the US) or in a coexistence of a national organic label and private labels (as is the case in most of Europe). Both approaches have specific opportunities and risks, mostly related to the ability of farmer associations that implement more stringent regulations, to communicate them (see Boström and Klintman, 2006). However, in Thailand there is no national label being used on every organic product, with an optional private label. Rather there are two separate labels, adding to the confusion caused by the plethora of labelled food. Thus, establishing a national label could help establish a profile for organic products and communicate a clear message to consumers.

Two approaches could help the private sector to increase the market share of organic products: increasing the purchasing frequency of the 'organic buyers' and encouraging the 'organic non-buyers' to try organic products. Increasing the purchasing frequency of those who are aware of organic products and who purchase them for health and environmental reasons could be achieved by improving availability, offering a wider range of organic vegetables and ensuring that they are always very fresh. Freshness may be a reason for those who purchase organic products for hedonic reasons to become regular buyers (see Sakagami *et al.*, 2006).

Consumers should also be provided with additional information on organic farming, so that they may, e.g. in case of a food scare, be aware that organic production methods and organic certification provides additional benefits that may address their concerns. Providing consumers with information on organic farming and the distinctiveness of organic food compared to 'safe food', may encourage the 'organic non-buyers' to purchase organic products. Although it might not allow the capturing of those consumers who are satisfied with 'safe foods' and have no need for additional benefits, it might allow producers and retailers to convince some of those consumers who are simply unaware of organic farming. By emphasizing that organic products are produced entirely without synthetic pesticides and are thus least likely to have residues, producers and retailers would address the health aspect, which is the most important issue to consumers. If this message can be conveyed to consumers, it would give organic products a unique and distinctive position on the market and overcome the current lack of distinction between organic, 'safe' and 'hygienic' products.

To convey this message and to ensure its credibility, consumers must be able to trust the control system based on independent, accredited certification agencies and annual inspections based on clear standards for organic farming. As Kramol *et al.* (2006) report, the majority of consumers in Northern Thailand would prefer to purchase products certified by accredited agencies. Being certified by a certification agency that is accredited by an international body (e.g. IFOAM) could increase the trust of consumers towards the certification process, as they can rely on the fact that it complies with international standards (see Barrett *et al.*, 2002).

Increasing consumers' awareness of organic farming, their trust in the rigorous inspections and the organic certification system as well as increasing the availability and range of fresh organic vegetables, may be the most effective way of increasing their market share. Overall, organic foods producers and retailers are well placed to answer the Thai consumers' health concerns, if they can convey their organic claims with conviction and thus distinguish themselves from other 'safe food' labels.

Discussion in lecture room

Apply your theoretical knowledge with respect to this paper:
- The principal component analysis resulted in three dimensions: health & environmentally friendly, fun & fresh and tasty. In the fun & fresh category one statement was 'I just wanted to try something new'. Consumers looking for new products and regularly switching between brands are called 'variety seekers'. Search for studies on the share of variety seekers for different food categories such as dairy products, vegetables, bread. Discuss the findings.
- This paper reports price differences between conventional and organic vegetables in Bangkok of between 100-170%. Compare these price differences with the average price differences reported for Europe (see Haas *et al.*, 2010, *Organic food marketing in the European Union; a marketing analysis*) and explain the differences from the aspect of supply and demand and the stage of market development (life cycle theory).
- In Thailand many competing 'safe food' labels contribute to consumer confusion. Food in general is considered to be a low-involvement product. Read the literature regarding communication for low-involvement products. Based on these insights, develop a communication strategy to clearly differentiate a national organic Thai label from all the other competing labels.

Acknowledgements

We would like to thank the students of Thammasat University who participated in the study and interviewed the consumers. Special thanks go to Nakorn Limpacuptathavorn for his tireless support. We are grateful to the managers of the supermarkets and health food shops for allowing us to survey their customers. We also thank Vitoon Panyakul, Kaan Ritkhachorn, Arada Gongvatana and Prinya Pornsirichaiwatana for offering us their insights into the organic food market in Bangkok. However, the opinions expressed here and all errors and omissions are our own. We are also grateful to the reviewers for their detailed and helpful suggestions.

References

ACNielsen, 2005. Functional food and organic. A global ACNielsen online survey on consumer behaviour and attitudes. Available at: http://www2.acnielsen.com/ reports/documents/ 2005_cc_functional_organics.pdf. Accessed June 2009.

Baker, B., Benbrook, C., Groth, E. and Benbrook, K., 2002. Pesticide residues in conventional, integrated pest management (IPM)-grown and organic foods: insights from three US data sets. Food Additives and Contaminants 19: 427-446.

Barrett, H., Browne, A., Harris, P. and Cadoret, K., 2002. Organic certification and the UK market: organic imports from developing countries. Food Policy 27: 301-318.

Batte, M., Hooker, N., Haab, T. and Beaverson, J., 2007. Putting their money where their mouths are: Consumer willingness to pay for multi-ingredient processed organic food products. Food Policy 32: 145-159.

Boström, M. and Klintman, M., 2006. State-centered versus nonstate-driven organic food standardization: A comparison of the US and Sweden. Agriculture and Human Values 23: 163-180.

Buzby, J., 2001. Effects of food-safety perceptions on food demand and global trade. In: Regmi, A. (ed.), Changing structure of global food consumption and trade. Agriculture and Trade Report WRS-01-1. Washington: Economic Research Service, US Dept. of Agriculture, pp. 55-66.

Chaivimol, S., 2003. Marketing green and organic agricultural produce in Thailand. Paper presented at the Expert Group Meeting Marketing green and organic agricultural produces as a tool for rural poverty alleviation, held 27-31 October 2003 in Changyuan County, People Republic of China.

Chunyanuwat, P., 2005. Country report Thailand. In: Proceedings of the Asia Regional Workshop: Implementation, monitoring and observance – International code of conduct on the distribution and use of pesticides. RAP Publication 2005/29. Bangkok: FAO Regional Office for Asia and the Pacific.

Delforge, I., 2004. Thailand: the world's kitchen. Le Monde Diplomatique, July 2004. Available at: http://mondediplo.com/2004/07/05thailand. Accessed June 2009.

Eddleston, M., Karallieddec, L., Buckley, N., Fernando, R., Hutchinson, G., Isbister, G., Konradsen, F., Murray, D., Piola, J.C., Senanayake, N., Sheriff, R., Singh, S., Siwach, B. and Smit, L., 2002. Pesticide poisoning in the developing world – a minimum pesticides list. The Lancet 360: 1163-1167.

Eischen, E., Prasertsri, P., Sirikeratikul, S., 2006. Thailand's organic outlook. Global Agriculture Information Network Report, USDA Foreign Agricultural Service. Available at: http://www.fas.usda.gov/ gainfiles/200009/30678084. pdf. Accessed June 2009.

Fillion, L. and Arazi, S., 2002. Does organic food taste better? A claim substantiation approach. Nutrition and Food Science 32: 153-157.

Foodmarket, 2003. Shrimp: A review of the news in 2002. Availble at: http://www.foodmarket exchange.com/ datacenter/industry/article/idf_shrimp_review2002.php. Accessed June 2009.

Grunert, K., 2005. Food quality and safety: consumer perception and demand. European Review of Agricultural Economics 32: 369-391.

Haas, R., Canavari, M., Pöchtrager, S., Centonze, R. and Nigro, G., 2010. Organic food in the European Union: a marketing analysis. In: R. Haas, M., Canavari, B. Slee, C. Tong and B. Anurugsa (eds.), Looking east looking west: organic and quality food marketing in Asia and Europe. Wageningen Academic Publishers, Wageningen, the Netherlands, pp. 21-46.

Hamm, U. and Gronefeld, F., 2004. The European market for organic food: Revised and updated analysis. Organic Marketing Initiatives and Rural Development, Vol. 5. University of Wales, Aberystwyth, UK.

Hardeweg, B. and Waibel, H., 2002. economic and environmental performance of alternative vegetable production systems in Thailand. Paper presented at the international symposium 'Sustaining food security and managing natural resources in Southeast Asia – Challenges for the 21st century', held 8-11 January 2002 in Chiang Mai, Thailand.

Hoogland, C., De Boer, J. and Boersema, J., 2007. Food and sustainability: Do consumers recognize, understand and value on-package information on production standards? Appetite: 47-57.

IPM-DANIDA, 2003. Did you take your poison today? A report by the IPM DANIDA project: 'Strengthening Farmers' IPM in pesticide intensive areas'. Available at: http://www.ipmthailand.org/en/. Accessed June 2009.

IPM-DANIDA, 2004. Pesticides-health surveys. Data from 606 farmers in Thailand. Report 62 by the IPM DANIDA project: 'Strengthening Farmers' IPM in pesticide intensive areas'. Available at: http://www.ipmthailand.org/en/. Accessed June 2009.

Jaffe, J. and Gertler, M., 2006. Victual vicissitudes: Consumer deskilling and the (gendered) transformation of food systems. Agriculture and Human Values 23: 143-162.

Kosulwat, V., 2002. The nutrition and health transition in Thailand. Public Health Nutrition 5: 183-189.

Kramol, P., Thong-ngam K., Gypmantasiri, P. and Davies, W., 2006. Challenges in developing pesticide-free and organic vegetable markets and farming systems for smallholder farmers in North Thailand. Acta Horticulturae 699: 243-251.

Lohr, L., 2001. Factors affecting international demand and trade in organic food products. In: A. Regmi (ed.), Changing structure of global food consumption and trade. Agriculture and Trade Report WRS-01-1. Washington: Economic Research Service, US Dept. of Agriculture, pp. 67-79.

Lockie, S., Lyons, K., Lawrence, G. and Mummery, K., 2002. Eating 'green': Motivations behind organic food consumption in Australia. Sociologia Ruralis 42: 23-40.

Lotter, D., 2003. Organic agriculture. Journal of Sustainable Agriculture 21: 59-128.

Magkos, F., Arvaniti, F. and Zampelas, A., 2006. Organic food: Buying more safety or just peace of mind? A critical review of the literature. Critical Reviews in Food Science and Nutrition 46: 23-56.

Magnusson, M., Arvola, A., Koivisto Hursti U.-K., Åberg, L. and Sjödén, P.-O., 2003. Choice of organic foods is related to perceived consequences for human health and to environmentally friendly behaviour. Appetite 40: 109-117.

McEachern, M. and McClean, P., 2002. Organic purchasing motivations and attitudes: are they ethical? International Journal of Consumer Studies 26: 85-92.

Moen, D., 1997. The Japanese organic farming movement: Consumers and farmers united. Bulletin of Concerned Asian Scholars 29: 14-22.

Nelson, J., 1991. Marketing of pesticide-free vegetables in Bangkok. MSc Thesis at the Asian Institute of Technology, Bangkok, Thailand.

Padel, S. and Foster, C., 2005. Exploring the gap between attitudes and behaviour. Understanding why consumers buy or do not buy organic food. British Food Journal 107: 606-625.

Panyakul, V., 2002. National Study: Thailand. In: Organic agriculture and rural poverty alleviation Potential and best practices in Asia. United Nations Economic and Social Commission for Asia and the Pacific (UNESCAP), pp. 173-203.

Pénau, S., Hoehn, E., Roth, H.-R., Escher, F. and Nuessli, J., 2006. Importance and consumer perception of freshness of apples. Food Quality and Preference 17: 9-19.

Posri, W, Shankar, B. and Chadbunchachai, S., 2007. Consumer attitudes towards and willingness to pay for pesticide residue limit compliant 'safe' vegetables in Northeast Thailand. Journal of International Food and Agribusiness Marketing 19: 81-101.

Roitner-Schobesberger, B., 2006. Consumers' perception of organic foods in Bangkok, Thailand. MSc Thesis at the University of Natural Resources and Applied Life Sciences, Vienna. Available at: http://www.wiso.boku.ac.at/2561.html. Accessed July 2009.

Sakagami, M., Sato, M. and Ueta, K., 2006. Measuring consumer preferences regarding organic labelling and the JAS label in particular. New Zealand Journal of Agricultural Research 49: 247-254.

Srithamma, S., Vithayarungruangsri, J. and Posayananda, T., 2005. Food safety programme: A key component for health promotion. Available at: http://www.fda.moph.go.th/ project/foodsafety/HealthPromotion2.pdf. Accessed July 2009.

Thompson, G., 1998. Consumer demand for organic foods: what we know and what we need to know. American Journal of Agricultural Economics 80: 1113-1118.

Tiensin, T., Chaitaweesub, P., Songserm, T., Chaisingh, A., Hoonsuwan, W., Buranathai, C., Parakamawongsa, T., Premashthira, S., Amonsin, A., Gilbert, M., Nielen, M. and Stegeman, A., 2005. Highly pathogenic avian influenza H5N1, Thailand, 2004. Emerging Infectious Diseases 11: 1664-1672.

UNDP, 2007. Thailand human development report 2007 – Sufficiency economy and human development. United Nations Development Programme, Bangkok, Thailand.

Veeck, A. and Veeck, G., 2000. Consumer segmentation and changing food purchase patterns in Nanjing, PRC. World Development 28: 457-471.

Wandel, M. and Bugge, A., 1997. Environmental concerns in consumer evaluation of food quality. Food Quality and Preferences 8: 19-26.

WHO, 2006. The WHO recommended classification of pesticides by hazard and guidelines to classification 2004. World Health Organization, Geneva, Switzerland.

Yiridoe, E., Bonti-Ankomah, S. and Ralph, C., 2005. Comparison of consumer perceptions and preference toward organic versus conventionally produced foods: A review and update of the literature. Renewable Agriculture and Food Systems 20: 193-205.

Zanoli, R., 2004. The European consumer and organic food. Organic Marketing Initiatives and Rural Development Vol. 4.: School of Management and Business, Aberystwyth, UK.

Zhang, X., 2005. Chinese consumers' concerns about food safety: Case of Tianjin. Journal of International Food and Agribusiness Marketing 17: 57-69.

Zhao, X., Chambers, E., Matta, Z., Loughin, T. and Carey, E., 2007. Consumer sensory analysis of organically and conventionally grown vegetables. Journal of Food Science 72: 87-91.

Quality, organic and unique food production and marketing: impacts on rural development in the European Union

B. Slee

Socio-Economics Research Group, Macaulay Institute, Craigiebuckler, Aberdeen AB15 8QH, United Kingdom

Abstract

This paper reviews the nature of European food markets and the place of food differentiated by quality standard in that market. Some aspects of the production of 'alternative' 'differentiated' and 'speciality' foods are likely to impact positively on rural livelihoods but the current evidence base is weak. The presumed linkages between speciality and alternative food production and rural development are rarely quantified. The European Union has developed policies to protect regionally specific and organic foods, but over and above the EU standards there are many other private sector and national and local government means used in promoting regionally produced food. Alternative Agriculture and Food Networks (AAFNs) may provide a potentially useful means of supporting rural development, but if demand is fickle, the development base may be fragile. As an alternative to the mainstream food system AAFNs tend to be more labour intensive and are less vulnerable to cheap commodity imports than mainstream production, but their inherent fragility should not be ignored.

1. Introduction

In 2006 the European Farm Commissioner Mariann Fischer Boel asserted the need to:
> 'conduct quantitative research into the economics of food quality schemes, especially covering the impact on farm income and rural development, and have discussion on the redistribution of money between the farmer and the retailer.'

This paper explores the potential of quality, organic and unique foods to contribute to rural development in the European Union, recognising, along with the European Union Farm Commissioner, the limited amount of information currently available. First, it briefly reviews the nature of European food markets and the place within these of food where either origin of production or a certified production process and/or breed creates food products differentiated from commodity foods. Second, it identifies the features of the production systems and supply chains which are likely to impact positively on rural livelihoods and rural development. Third, it explores the policy context and range of manifestations of place-, production-, process- or breed-certified systems of production within the European Union. Finally, it explores the potential alternative futures for such systems in contributing to wider rural development objectives.

In mature markets, product differentiation is a standard means of trying to gain or retain market share. Differentiation of a food product can arise from branding as a commercial consideration or by the policy-driven protection of the identity of a product because of a distinctive production process or place of production. Often the policy-driven identification of a distinctive production process protects artisanal and traditional methods against mass-produced modern alternatives. Sometimes the protected product relates to a particular breed or variety. The protection of regional identity and special production processes is part of the regulatory system that shapes a significant part of European food production, distribution and consumption.

At a broader scale, regulation in the food sector in Europe stems from two main concerns. First, there has been much pressure to produce 'safe' food; that is, food free from chemical or microbial contamination or food products for which the constituents, including major ingredients, preservatives etc. must be clearly specified. In the wake of various food 'scandals' around the world, including within the European Union, there has been an enormous investment in certification systems which guarantee that food is safe to eat. The scares relating to toxins in milk in Belgium, contaminated olive oil in Spain and the epidemic of Bovine Spongiform Encephalopathy in Britain and other European countries brought food safety issues to the fore in the 1990s and underpinned the development of more rigorous testing regimes and traceability procedures. Frequently these are business to business schemes, which increasingly enable traceability. They are often virtually invisible to the final consumer but are rightly perceived as vital to the management of safe food supply chains between producer and retailer. Important though these concerns are they are not the subject of this paper.

Secondly, alongside the demand for safe food, there has been a strong desire from some parts of the European Union to protect by official designation distinctive food production systems and regional food products and use these designations as a platform for protecting the producer. As well as geographical specificity, specificity can also be asserted by the production system (such as organic or biodynamic) or by breed (e.g. Aberdeen Angus or Hereford beef). These schemes are built around a desire to enhance product information from business to consumer but also have the capacity to build and share knowledge amongst groups of producers. They entail another approach to differentiation, driven more by national or regional government rather than the mainstream food producer or retailer.

In the international food and beverage industry, the dominant form of differentiation has long been by brand. Behind a particular brand lie in-house procedures for ensuring that the brand identity is preserved by the application of consistent methods of production and the use of marketing strategies, as well as protection of that brand by law. Companies go to enormous effort to nurture and preserve the power of the brand, not least because brand leadership is often associated with high levels of profitability.

Not only is branding a strategy for the commodity producer trying to differentiate his/her product; it is also widely used in relation to regional products. In their marketing strategies, some companies often try to assert their connection to a particular locale and the brand is associated with a particular environment. Malt whisky provides an excellent example of this process of identifying a generic product with a particular place which has particular environmental qualities. Where the distinctive qualities of space are embodied in the product, and the product has international reach in food markets, the impact on rural development can be considerable. Essentially, a brand which has such regional associations is an assertion of purported link between product distinctiveness and place. The different commercial brands of malt whisky are using essentially the same logic of place connection that is found in regional branding schemes, but refining down the spirit of the product to particular locales, rather than whole regions and linking it to specific locales.

In this paper, however, we concentrate on certification processes related to the distinctiveness of a production method, processing method or breed and how the regulatory structures associated with these create opportunities for enhanced rural economic well-being. We recognise that there are other layers of regulation, some relating to food safety, others to protecting brands and others to protecting the producers in a particular place, of a particular product or of a particular breed. Whilst certain similarities exist in any form of certification, notably the desire to separate out one type of product from the commodity norms, we are not concerned with the rural development impacts of commercial branding, or of wide-ranging 'farm quality assured' types of certification. Our focus of interest is the measures that relate to quality certification relating to places of production, types of process or specific breeds or varieties that are normally associated with particular regions.

Rural development is broadly understood to comprise endogenously or exogenously induced action by private, public or voluntary sector actors that enhances the well-being of rural people. The development outcome might be measured in terms of economic gains, but other types of benefit such as gains in health, cultural capital, human capital and empowerment of rural people are also widely considered to constitute rural development.

It is recognised that *sustainable* rural development must be built around particular principles and that these principles are usually considered to have socio-cultural, economic and environmental dimensions (WCED 1987). It is also recognised that the very notion of sustainability is widely contested and that different individuals, groups and stakeholders may value the different dimensions of sustainability differently. In developing countries, not least because of the emphasis on improving self-sufficiency and supporting sustainable environmental management, the term sustainable livelihoods is widely used almost synonymously with rural development. The livelihoods approach emphasises that gains in wellbeing can arise from endogenously or exogenously induced positive changes in human, financial, environmental, social or physical capital.

A central issue with region-specific or production process-specific foods is what characterises 'different-ness' in the food product in question. At European level, a number of geographical origin labelling designations are now in place and have absorbed many predecessor national schemes (see below). However, from a consumer perspective, the simple designation of a product as local or regional may well resonate with consumers and effectively over-ride official designations and regulations. Indeed, market research surveys in the UK have shown the frequent confusion of consumers between local food and organic food. Thus we cannot encapsulate the whole market for regionally specific foods within the regulated and defined demarcations. We also need to recognise that many consumers will buy locally produced food for its perceived benefits of freshness or 'localness' regardless of formal designation.

From a rural development perspective, a demand for local or regional food, whether designated or not, can provide a compelling and primary reason for supporting a local food system in which the informal designation of local as a product attribute may override other more formal designations. A recent article in Time magazine (Cloud, 2007) reviewed the organic versus local dilemma, which has been further explored by Pretty and others (2005). In the UK study, Pretty *et al.* conclude that local foods had, on average, a lower environmental footprint that the available organic basket of food. Oglethorpe (2007) has recently challenged their conclusions, arguing that in general a conventional supermarket based food system produces a lower environmental footprint than the local food sector. Tregear *et al.* (2007) argue that certified food systems have potential to support rural development.

In practice, localness as an attribute is often combined with other designations, in that the pioneers of local food are often involved in specific production practices associated with designated foods. Further, bodies promoting organic food often also advise consumers to buy *local* and *organic*. Local foods are often perceived as integral elements of local culture. Eating locally produced food is often an assertion of local identity and a cultural act that indicates a desire to sustain local food cultures. A further value of localness lies in the potential for reduced food miles and the possible lowering of the carbon footprint of local food.

2. The nature of European food markets

It is extremely difficult to generalise about the nature of European food markets, although some broader regional groupings of market characteristics are discernible. In general, southern/ Mediterranean European countries have much less concentrated food retail sectors, but can still exhibit high levels of concentration in the processing sectors. In the Mediterranean region, there is a much stronger representation of regional food in people's food choices and this is backed by high levels of regional and even more local identity production. In contrast, most Northern European countries have rather high levels of both retail and processing sector dominance. Food products are highly differentiated but more by brand than place attribute. The retail end of the market is generally seen as the hub of power in the food supply chain and this tends to drive a food system where homogeneity of production is sought by the retailer and tight product specifications favour industrial style production processes. In new member states in Central and Eastern Europe, some elements of the food industry were rapidly penetrated by foreign direct investment and the current food sectors are a mixture of the legacy of communist period state industries, new post-communist processing and a strong micro-scale of direct marketing and domestic production and self-consumption of fruit and vegetables.

The same general drivers of demand operate throughout Europe. Until recently, the general increase in incomes had been a major driver of demand. In the uncertain economic climate of 2008, the so-called 'credit crunch' had a certain impact on food demand, both in relation to routine and luxury items and eating out. Budget shops and budget brands, especially private labels, appear to be prospering in the uncertain market conditions as the global economy drifts towards recession. The extraordinary turbulence in global financial markets in 2008 has created uncertain economic conditions which few have experienced in their lifetimes to date and which have brought repercussions in all sectors of the economy. The collapse of major banks, the tightening of lending and the drift of western economies towards recession are having enormous impacts on consumer spending. Contemporaneously with this trend but completely unconnected to it, global food markets have been as tight as at any time in the previous 20-30 years and primary commodity prices rose dramatically in 2007. The annual rate of inflation in food prices has generally exceeded the retail price index. It seems probable in such a financial climate that luxury or niche items in the shopping basket will be particularly affected. Early reports from the UK suggest that cash-strapped consumers may also be turning away from organic produce because of its high price compared to other conventional foods with a report that spending on organic food and drinks fell from a peak of nearly £100m a month earlier this year to £81m in the most recent four-week period recorded (Jowitt, 2008).

Many other factors have affected the changed demand for food in Europe. The tendency, especially in more urbanised areas, for more women to work outside the home has tended to reduce the available time for food preparation and increase the demand for either ready meals or pre-prepared ingredients. Demands for traceability have grown in the wake of food scares and elaborate traceability systems are now in place. Travel has opened up tourists to new food experiences. A growth of multiculturalism has crated highly diverse readily available cuisines both in food stores and restaurants.

Over time and for a variety of reasons, there has been a degree of alienation of some consumers from the highly concentrated elements of the food system. A small proportion of food consumers has articulated an alternative vision for food, which yearns for local differentiated production and is associated with a search for authenticity, traceability and distinctiveness in food. This Europe-wide movement has similarities to the extant legacy of regional food systems of southern Europe, some of which have extended their market reach, whilst other products have remained trapped within regional market networks. In North America the idea of a foodshed has been asserted as a catchment area for local food and there are wider discussions of the desire for local food in the context of bioregionalism (McGinnis 1999).

At farm level, product prices have often been depressed by powerful food chain operators and the farm-retail price spread has widened. More and more of the retail value of food has gone to processor or retailer and less to the farmer. This has been a further factor incentivising some farmers to work outside the mainstream commodity chains. Relocalising production and adding value to unprocessed food by either processing or direct marketing have emerged as significant adjustment strategies by the farming community (Slee, 1989). At the same time, in southern parts of Europe, regional production systems have survived more because of residual local production practices and strong consumer demand than by explicit adjustment strategies.

Most food retail markets have shown increased concentration of ownership in the last 20-30 years. Food retailing has generally been associated with an eclipsing of local brands but a contemporaneous proliferation of brands by producers and retailers using their own market power to develop and promote their own brands which often ignore local provenance. Thus the contemporary consumer is confronted in-store by a remarkable number of product lines, which are differentiated by marketing strategies. Some supermarkets now sell as many as three grades of own-branded produce for the same product: the basic food good for budget shoppers; the standard line; and the special product in order to reach out to different shoppers within the same store.

The dominance of the major food retailers is widely evident but geographically variable. It is greatest in Nordic and in northern and western Europe and lower in the south of Europe. Particularly in southern Europe, there is a strong legacy of regional foods and the reach of international commodity-based foods is weaker. This legacy of regionally distinctive and often artisanal food is often now protected by regulation, formerly at national level, now at European level.

In spite of that dominance of major retailers and food producers in more northern European food markets, there is a small but dynamic component of the food market that operates outside the mainstream retailers. This is often described as the alternative food sector or as Alternative Agro-Food Networks (AAFNs) (Goodman, 2004). This has been the subject of a great deal of attention in spite of its modest size. Those who have explored its characteristics regard it as a potential harbinger of future food system structures in Europe (O'Connor *et al.*, 2006). There are similar characteristics in food markets in most developed western economies, with local production and regional identity products assuming an important role in the promotion of local food in the retail and food service sectors.

Slee and Kirwan (2007) suggest that this polarised distinction between the two sectors oversimplifies the actual situation which contains many features of hybridity. Not only are there hybrid forms but the linkages between the mainstream and alternative sectors are constantly mutating and evolving. The organic sector which represents an early form of alternative food, operating largely outside the market place in its early years of development in the 1950s has been partly 'conventionalised' in the opinion of some commentators (Lockie and Halpin 2005).

The generalised pattern of demand for food in Europe thus shows a structural shift towards large scale supermarket retailers, who stock a huge product range. The generalised pattern of supply has responded to the demands of these dominant players.

3.The linkages between certified food and rural development

There is widespread evidence that as economic development proceeds, the economic importance of food in consumer choices declines. Many commodity foodstuffs have income-inelastic demand (Tomek and Robinson, 1990). In the case of some so-called 'inferior' products, demand may even contract as incomes rise. This means that general economic advance will normally turn the terms of trade against food producing interests. However, the continued growth of the food retail sector at a time of declining retail expenditure on food remains something of a mystery, although economies in distribution and an ability of powerful retailers to suppress prices to farmers or processors have probably contributed to this. So too has the tendency for many major supermarkets that began as predominantly food retailers to expand into a much wider range of consumer goods and services and use their established distributional efficiencies on a wider range of products.

Over the last decade, there has been a long-running debate about 'freeing up' trade in food products. This is part of a neo-liberal trend in global economic policy, in which food trade is a relative latecomer. Almost all developed countries, certainly those of the old world, have had high levels of protection of their own farmers in the relatively recent past, but this has been challenged by the World Trade Organisation and its predecessor body, the General Agreements on Tariffs and Trade. National level protectionism has been challenged and largely vanquished by the drawing of agricultural trade into the remit of the WTO over the 1990s. The reassertion of place identity and the development of policy to support distinctive production systems can be seen at one level as an attempt to protect particular forms of production, either by creating enhanced conditions for differentiation or by appealing to the demands of particular groups of consumers. The emergence of support for unique and quality foods has grown in part at least out of the particularly strong resistance to globalisation in some parts of European agro-food systems.

The neo-liberal model of rural development eschews regional, national or federated states boundaries. According to its underlying tenets, allocative efficiency will be best achieved when there are least barriers to trade. Accordingly, low cost regimes will gain advantage from the reduction in barriers to trade and enhanced rural development will be the outcome. There are of course many impediments to such an outcome. Powerful vested interests of international traders, unstable and corrupt political regimes and an inability to factor into such a model the environmental damage caused by low-cost production raise many doubts about the achievement of rural development outcomes where they are most needed.

Interest in rural development has grown as the evident failure of a modernised agriculture-dominated rural economy to deliver adequate economic wellbeing to the rural population has become apparent. It has long been argued that rural development (at least in its developed country form) is weakly theorised, although there is some evidence of theoretical developments in regional economics and regional geography which inform rural development theory. Many contemporary theories revolve around the identification of social capital and networks which are seen to provide the collaborative platform and trust on which new development networks are based. Thus Camagni's work on the innovative milieu (Camagni, 1995) and Storper's work on embeddedness (Storper, 1995) both posit the existence of networks, untraded interdependencies and trust between actors.

Few of these contemporary theorisations of rural development in developed countries posit a central role for the farm sector. The work of the Wageningen rural sociologists and their collaborators is an exception. There has been much concern regarding the effects of the so-called agricultural modernisation project (Marsden, 2006; Van der Ploeg, 2003) on European rural development. The policy-assisted modernisation of agriculture since 1945 has generally been associated with a substitution of capital for labour with improved mechanisation and a decline in the agricultural workforce (REF). The net result has sometimes been what Wibberley (1981) has described as 'strong agricultures and weak rural economies', areas where the modernisation of the agriculture has strengthened agribusiness, but weakened the relative importance of agriculture, and, at the same time, reduced the capacity for what the OECD (1999) has more lately termed 'cultivating rural amenities'. The support for a technocratic modernisation project has often replaced locally specific production methods by core technologies and reduced agro-biodiversity.

The Wageningen group, building on work on endogenous development theorised by Van der Ploeg and Long (O' Connor 2006; Van der Ploeg and Long 1994), have argued that the post Second World War agricultural modernisation project in Western Europe has compromised both farm incomes and the environment and that both might be better served by more regional or local modes of production and new networks of food production and consumption. Their model is explicitly European. They argue for a new model of rural development based on the relocalisation of food systems, the diversification of farming and the re-engagement with endogenous production approaches. Where the former endogenous models of production have largely been eclipsed by the modernisation project, the term neo-endogenous development has been coined (Ray, 2006).

The Wageningen model is built on the identification on the transformation of the ago-food sector from an apparently imploding model based on the modernisation paradigm to an alternative system built around Alternative Agriculture and Food Networks (AAFNs). There is, however, at present no conclusive evidence on the rural development impacts of AAFNs. The European Commission in its green paper acknowledges that 'There is still a lack of data on the value adding process in Food Quality Certification Schemes and their impact on rural development' (DG Agri, 2008).

The optimistic view of AAFNs has been challenged by North American commentators such as Goodman (2004) as presenting an unrealistically optimistic view of the ability of the alternative farming and food networks to challenge agribusiness. Goodman argues that the continued dominance of large scale retailers and processors attests to the adaptive capacity of multi-national retailers and producers. Further the sum of the impacts of these large scale production, processing and distribution systems undoubtedly impacts on rural development. Their decline or demise would cause a considerable loss of employment and income in many specialised food production regions.

The OECD perspective on a new paradigm for rural development is broadly complementary to the Wageningen approach in that it too argues that rural development may be better engendered by new modes of organisation (OECD, 2006), but stresses that the new approach needs to be rooted in new modes of governance and new policies which give much greater emphasis to non-farm rural development than in the Wageningen model which places rather more emphasis on the farm as the principal unit of delivery of both new food and non-food products and services.

There are a number of reasons for asserting a link between certified food and rural development. These include, firstly, the economic advantages arising from a protected label for foods of a distinctive nature, which prohibits their production by those who do not comply with the regulatory framework. Second, the production processes involved in the production of many designated foods may be based more on artisanship and less on mass production, thereby creating more rural employment. Third, the food supply chains operated by many but by no means all designated food producers are often local by intent and this can encourage shorter food supply chains and result in farmers retaining a higher proportion of the retail pound spent on food. Fourth, the designated products may often be worth more per unit volume, thus rewarding primary producers more and resulting in greater direct, indirect and induced benefits in rural economies than might be obtained for mainstream products.

The principal economic advantage arising for a farmer or food producer involved in a certificated production process is the general ability to limit production to a restricted group of producers. This creates a quasi-monopoly, the strength of which is contingent on the similarity of the product to others from similar areas. As long as the certificated artisanal product in question is differentiated and seen as superior in quality to the commodity-based alternative, there is, however, a strong likelihood that the producer will receive a better price.

A significant debate has arisen about the role of what is termed 'conventionalisation' of Alternative Agro-Food Networks. This process of conventionalisation occurs when a form of production that has developed as alternative is drawn into the mainstream agro-food system. This has happened to a degree with organic food in most western countries, in that a very large proportion of organic food is purchased through supermarkets and reaches the final consumer through mainstream food distribution channels. Such marketing arrangements potentially make the alternative food producer or processor equally vulnerable as the commodity food producer. Further the impacts of what may be long-distance movement of food on agro-food system sustainability raises doubts about the aggregate benefits of such distribution channels.

It remains uncertain and untested as to whether conventionalised alternative food systems create greater benefits for rural development than do mainstream conventional production systems. In some cases the conventionalised alternative may generate higher margins to farmers, even if the product is purchased by a powerful retailer with the market muscle to drive prices down. However, some large scale retailers may enter arrangements with consortia of farmers to improve product quality in integrated supply chain management initiatives which also have the potential to enhance farmer returns.

4. Policy and practice in European quality and designated food markets

There are a number of policy drivers in protecting certain types of product and production in European agriculture, or indeed protecting any distinctive production system or product. A major European motive in policy support for distinctive products in Europe is the desire to nurture and support the European model of agriculture based on the idea of multifunctionality (OECD, 2004). There are assumed synergistic effects of traditional food production systems on multifunctionality, with potential benefits in relation to rural culture, biodiversity and landscapes.

The European Union has standardised a range of national level schemes which protect regionally specific foods and organic food. The recent Green Paper (Directorate General Agriculture and Rural Development, 2008) identifies four types of Europe-wide protection:
1. protected denomination of origin (PDO); protected geographical origin (PGI) and the GI (geographical indication) schemes proposed in the wines and spirits sectors;
2. traditional speciality guaranteed (TSG);

3. products from the outermost regions; and
4. organic agriculture.

Alongside these schemes, a raft of other measures exist to recognise particular production standards and to meet supermarket and other food chain actors' demands for quality assurance.

A second reason to support quality, unique and organic systems is the potential benefits they provide for rural development. First, the protection of a unique food product or a regional product confers advantage on that region. Second, the characteristics of the protected system may create beneficial rural development outcomes. Employment may be more intensive on such foods. They may offer greater scope for direct marketing thus ensuring that the farm-retail price spread is less and exploitative arrangements by powerful retailers are avoided. Further, the regionalisation of the production system may result in stronger regional/local linkages thereby enhancing the already advantageous employment effects. Additionally, the synergistic benefits on biodiversity may also deliver regional added value in that this may make the area more attractive as living space or for recreation or tourism. Given that the production systems associated with speciality foods are often less intensive there may be further benefits through reduced negative external effects, for example, in relation to greenhouse gas emissions or water pollution.

The LEADER territorial approach to rural development has often provided the partnership structure in which new regional food networks have been established and the mechanism for promoting and safeguarding old and developing new regional identity products. The LEADER approach has evolved since the early 1990s when it was introduced to provide support for an alternative development model suited to disadvantaged rural areas. Its widely perceived success has led to it being mainstreamed throughout rural areas of Europe in its latest incarnation.

There are many national and regional schemes, usually but not always with third-party certification procedures. Many formerly nationally designated foods have been drawn into the two dominant European Schemes: PDO and PGI. By 2006 there were over 700 such designated products in Europe (Table 1).

There is a very marked southern European bias in the frequency of these schemes by country and very weak engagement from the Nordic and North West European countries. The UK is something of an exception to the low engagement and after a somewhat condescending attitude early in the development of such schemes has now embraced them more vigorously. Germany too has made wide use of the designations. Broadly, this geographical variation represents the extent of regionally specific products, with a clear suggestion that distinctive products are more likely in agriculturally less favoured areas and areas with a strong tradition of regional food systems.

5. Potential alternative futures for certified quality food and rural development in Europe

The recent buoyancy of the market for differentiated food from the Alternative Food Supply Networks in Europe has been replicated in many developed countries. The effects of the recent economic downturn are unknown, but circumstantial evidence in the UK suggests it may be considerable. However, AAFNs are subject to essentially the same economic forces that impact on conventional food supply networks. A number of identifiable obstacles to growth can be identified.

Table 1. Number of PDO (protected denomination of origin), PGI (protected geographical indication) designations by country as of December 2006 (Giray, 2007).

Country	Number of schemes
Austria	12
Belgium	4
Czech Republic	3
Denmark	3
Finland	1
France	148
Germany	67
Greece	84
Ireland	3
Italy	155
Luxembourg	4
The Netherlands	6
Portugal	93
Spain	97
Sweden	2
United Kingdom	29
Total	711

The sheer proliferation of new enterprises within the alternative food sector ought to be a cause of concern. It can be argued that a critical mass of suppliers will enable a step change in AAFNs, with possible scale economies in distribution from local food hubs, which may be either ephemeral markets or permanent distribution depots. Alternatively, it might be argued that the proliferation of new local food systems will result in business casualties, where inadequate business and marketing planning or unanticipated competition weeds out the weaker players in the market.

To some extent some actors in AAFNs are insulated from market forces by having wealth from other sources and in running food enterprises more as a hobby than as a main source of livelihoods. Such lifestyle businesses may increase the supply and spread the consumer pound more thinly.

Small businesses may face very considerable transaction costs in using the market. The tight regulation of food safety and additional rules imposed by buyers may increase the cost of small suppliers using formal marketing channels. This may further incentivise small scale producers to use short supply chains or even direct marketing to reach the final consumer. On the other hand, supermarket buyers are increasingly keen to tap into the local food sector as an innovation seedbed, and make offers to food producers when they observe successful products. However, this raises issues of dependency on a single market and the associated increases in regulatory costs.

Many artisanal food enterprises produce food which may vary in flavour and consistency over time, particularly on a seasonal basis. Some dairy products will vary in quality depending on whether stock is grazed on pastures or fed conserved fodder in animal houses. The drive by supermarkets to source homogenous products often stands as a formidable barrier to those whose product may be less homogeneous because of artisanal production methods or seasonal variations in the quality of the raw materials.

The greatest barrier to development of quality food markets is likely to be the major economic downturn which is affecting global markets as a result of turbulence in financial markets in 2008. A major recession could impact on the demand for luxury products and refocus consumer attention on budget brands. There are signs of a downturn in some types of quality food in the organic sector and it is highly likely that this will transfer through to the wider speciality food sector.

Given the period of substantial growth in the alternative food sector and the scope for new entrants to find a niche, it is not clear how the sector will react tot the current downturn and how resilient it is to contracting demand. Paradoxically, lifestyle businesses may be more resilient than those businesses that are dependent on borrowing and which are more highly geared.

A final factor which might impact on the speciality food sector is the rise in commodity markets in 2007, which may have led some farmers providing niche markets to return to commodity production. The major price rises in commodities have proved relatively short-lived, but there is a general expectation of a rise in commodity prices above those of the recent historic past before the 2007 price rises.

Two extremes of futures can be sketched out regarding the likely future pathways of development of AAFNs and the certified quality food sector. On the one hand, pressure on global food systems to meet demanding climate change targets alongside growing concerns for food security and biosecurity concerns could lead to a partial relocalisation of food markets. On the other, the powerful retailers in the mainstream food system will adapt their offer to accommodate many of the elements currently provided by AAFNs and morph into a hybrid structure. Both offer some promise for rural development.

The relocalisation model draws aspirationally on a number of discernible consumer trends. It recognises the desire of some (mostly rather affluent) consumers to feel a greater degree of trust in food production and the desire for a more personal form of relationship with producer and seller. It recognises the market value of fresh food and regional identity foods. It potentially underpins a degree of relocalisation with consequent benefits on rural development in such locales. These are often likely to be the places that have received in-migration particularly of the creative class. They are likely to be areas where tourists visit and value the local culture including its food. They are likely to be areas where there is a residue of production systems or a dynamic human capital base to revalorise local production.

The assimilation model suggests a less significant injection of income into rural areas. Instead, it recognises the extreme market power of the dominant supermarkets, which though their power varies form country to country, have become more significant players in the last 30 years throughout Europe. The model suggests that the supermarkets will remain major players and will use their market power to 'mop up' the local and speciality food suppliers and assimilate them into their own distribution networks. At the margin, their use of market power might be expected to drive prices to the speciality and quality food sector down.

6. Conclusions

Food production, processing and distribution are intimately connected with, but not always important to, rural development. The general decline in the importance of the primary sector and the rise of the service sector in rural areas has resulted in development trajectories of many rural areas being more contingent on commuting possibilities and rural amenities than their food producing potential. However, whilst food has been largely eclipsed as the principal driver of positive rural economic change, it remains important, deeply so in some regions.

The so-called modernisation project in the European agro-food system which has underpinned change since the 1960s has generated very uneven outcomes. While production practices were being intensified in core areas, and an associated agro-food complex became a major provider for Europe-wide distribution systems, in more marginal areas, agricultural production was often compromised. In many such areas, a process of *desertification* occurred which comprised people leaving such areas and initiating a downward spiral of development. This process of *desertification* has created weak markets but left a residue of local food production and processing practices which have provided a possible platform for specialised production.

For a variety of reasons, many of the areas which formerly experienced declining economic fortunes have undergone a revival. This too is uneven. Areas with high amenity values have been destinations for lifestyle in-migrants who often bring with them enhanced demand for local identity products. The presence of significant proportions of creative people linked to media, education and the arts has been seen as instrumental in creating new economic growth in rural areas (Florida, 2003; Wojan and McGranahan, 2006). However, the aggregate demand for high-quality certified food in a volatile and declining market is uncertain and the evidence of enormous buoyancy must be tempered by concern.

It remains to be seen whether new drivers of food choice, such as responding to climate change, will impact on local and quality food markets. To some extent the local knowledge underpinning traditional production and processing may substitute human endeavour for bought-in energy. Food miles may be shortened if regional food markets strengthen, but the evidence on the capacity of the local food sector to deliver food into a low carbon economy is open to question.

On balance, the effect to date of the alternative and quality food sector on rural development has been almost wholly positive. Local development policy actors have embraced the quality food sector, not least because of its capacity to connect to tourism and create greater synergies. This can be seen evidenced in wine and food tours in many regions and in the promotion of food towns. However, as recession bites, it is uncertain whether the buoyancy of the market of the last decade will continue. The depth and extent of any economic downturn must be factored in to any consideration of how this sector contributes to aggregate rural development.

An optimistic longer term view suggests that economic expansion can resume and the role of speciality and certified food might be expected to rise, thereby delivering benefits to rural development, subject to the proviso that market power is not increasingly concentrated in the hands of others in the food supply chain. Even if the adverse economic climate remains, it is not inconceivable that the demand for local and potentially speciality food will grow, driven by concerns such as climate change, traceability and by changing consumer preferences. This could impact beneficially on rural development.

Discussion in lecture room

Apply your theoretical knowledge with respect to this paper:
- In 1973 the secretary of agriculture of the USA, Earl L. Butz, announced: 'What we want out of agriculture is plenty of food and that's our drive now. We have experienced a 180-degree turn in the philosophy of our farm programs; ... we turned back to the new philosophy of expansion. And it makes sense'. Compare the European view of a multifunctional agriculture with the US paradigm of a highly efficient agriculture. Try to find historical, geographical and cultural reasons for the differing views of agriculture.
- Search for scientific papers about multifunctionality of agriculture from European and US-authors and compare their findings.
- Search for the council regulation (EC) No. 510/2006, which regulates Protected Designation of Origin and Protected Geographical Indication. Discuss the differences between these two concepts.
- Read the book the *The ethics of what we eat* (2007) from Peter Singer and Jim Mason and discuss the economic concept of externalities. Bring examples of positive and negative externalities of agriculture. Discuss if efficient agriculture is the same as sustainable agriculture. Discuss how free markets deal with social and environmental cost of intensive farming practices!

References

Camagni, R.P., 1995. The concept of 'innovative milieu' and its relevance for public policies in European lagging regions. Papers in Regional Science 74: 317-340.

Cloud J., 2007. Eating better than organic. Time magazine. Available at: http://www.time.com/time/magazine/article/0,9171,1595245-5,00.html. Accessed January 2009.

DG Agriculture and Rural Development, 2008. Background paper to the green paper on agricultural product quality: Food quality certification schemes (FQCS), European Commission : Brussels.

Florida R, 2003. The rise of the creative class: and how it's transforming work, leisure, community, and everyday life. Basic Books, New York, USA.

Giray, F., 2007, Types of schemes operating in the EU: Overlap or synergy? Paper presented at Conference on Food Quality Schemes, Brussels.

Goodman D, 2004. Rural Europe Redux? Reflections on Alternative Agro-Food Networks and Paradigm Change. Sociologia Ruralis 44: 3-16.

Jowit, J., 2008, Shoppers lose their taste for organic food. The Guardian. Available at: http://www.guardian.co.uk/environment/2008/aug/29/organics.food1. Accessed January 2009.

Lockie, S. and Halpin, D., 2005, The conventionalisation thesis reconsidered: Structural and ideological transformation of Australian organic agriculture. Sociologia Ruralis 45: 284-307.

Marsden, T., 2006. Pathways in the sociology of rural knowledge. In P. Cloke, T. Marsden and P. Mooney (eds.), Handbook of Rural Studies. Sage, London, UK, pp. 3-17.

McGinnis, M.V. (ed.), 1999. Bioregionalism. Routledge, London, UK.

OECD, 1999. Cultivating rural amenities: an economic development perspective. OECD: Paris, France.

O'Connor D., Renting H., Gorman, M. and Kinsella J., (eds.), 2006. Driving rural development in Europe: policy and practice in seven EU countries. Van Gorcum, Assen, the Netherlands.

Oglethorpe, D., 2008. What's so good about Local Food? Agricultural Economics Society Conference on Our food economy - Are we connected yet? 24 Jan 2008, London, UK.

Pretty J.N., Ball A.S., Lang, T, Morison J.I.L., 2005. Farm Costs and Food Miles: An Assessment of the Full Cost of the UK Weekly Food Basket. Food Policy 30: 1-20.

Ray, C., 2006. Neo-endogenous rural development in the EU. In: P. Cloke, T. Marsden and P. Mooney (eds.), Handbook of Rural Studies. Sage, London, UK, pp. 279-291.

Singer, P. and Mason, J., 2007. The ethics of what we eat: why our food choices matter. Rodale books, London, UK.

Slee, B., 1989. Alternative Farm Enterprises. 2nd Edition, Farming Press, Ipswich, UK.

Slee, B. and Kirwan, J., 2007. Exploring hybridity in food supply chains, paper presented at 105th EAAE Seminar on International Marketing and International Trade of Quality Food Products. Bologna, Italy, March 2007.

Storper, M., 1995. The resurgence of regional economies, ten years later: the region as a nexus of untraded interdependencies. European Urban and Regional Studies 2: 191-222.

Tomek, W.G. and Robinson, K.L, 1990. Agricultural Product Prices. Cornell University Press, Cornell, Ithica, NY, USA.

Tregear, A., Arfini, F., Belletti, G. and Marescotti, A., 2007. Regional Foods and Rural Development: The Role of Product Qualification. Journal of Rural Studies 23: 12-22.

WCED, 1987. Our Common Future. Oxford University Press, Oxford, UK.

Wibberley G., 1981. Strong agricultures but weak rural economies – the undue emphasis on agriculture in European rural development. European Review of Agricultural Economics 8: 155-170.

Wojan T. and McGranahan, D., 2006. Does the Rural Creative Class Fuel Rural Innovation? Paper to Edinburgh meeting of OECD, October 2006.

Van der Ploeg, J.D.; 2003. The Virtual Farmer. Royal Van Gorcum, Assen, the Netherlands.

Van der Ploeg, J.D. and Long, A. (eds.), 1994. Born from within: practice and perspectives of endogenous rural development. Van Gorcum, Assen, the Netherlands.

Authors and editors

Christoph Ameseder

Expertise: Research focus on food supply chains with regard to trust in b2b relationships and research on organic food markets.

Current position: PhD student at the Institute of Marketing & Innovation, Department of Economics and Social Sciences, University of Natural Resources and Applied Life Sciences, Vienna (BOKU), Austria.

Web page: http://www.wiso.boku.ac.at/mi_ameseder.html
E-mail: c.ameseder@googlemail.com

Bundit Anurugsa

Expertise: Soil Science. He holds a Ph.D. in Environmental Science awarded by the Georg-August-Universität Göttingen. Current research interests regard organic agriculture and land environmental management and control. His research focuses on topics such as soil fertility restoration and carbon sequestration.

Current position: Assistant Professor of Environmental Science, Head of the Department of Environmental Science, Faculty of Science and Technology, Thammasat University, Rangsit Campus, Pathumthani, Thailand.

E-mail: banurugsa@yahoo.com

Maurizio Canavari

Expertise: Agri-food Marketing. Current research interests regard topics in agri-food marketing and economics of quality in the agri-food chains, such as trust and quality assurance in food networks, e-business and e-commerce in food value chains, marketing of quality, organic, and unique food in European and Asian contexts, consumer behavior towards traditional food products, functional food, genetically modified food, and organic products.

Current position: Associate Professor of Agricultural Economics and Appraisal at the Department of Agricultural Economics and Engineering, lecturing 'Agri-food marketing' and 'Agribusiness marketing' at the Faculty of Agriculture, Alma Mater Studiorum - University of Bologna (UNIBO), Bologna, Italy.

Web page: http://www.unibo.it/docenti/maurizio.canavari
E-mail: maurizio.canavari@unibo.it

Nicola Cantore

Expertise: Environmental economics, agricultural economics. Current research interests regard climate change, mitigation, adaptation and sustainable farming production practices.

Current position: Research fellow at the Overseas Development Institute, London, UK. Secondary affiliation: Università Cattolica del Sacro Cuore, Milan, Italy.

Web page: http://www.odi.org.uk/about/staff/profile.asp?id=599
E-mail: n.cantore@odi.org.uk

Maria Angeles O. Catelo

Expertise: Research focus on contract farming and other market-linking institutions for smallholder livestock producers, livestock waste management and environmental impacts, industrial pollution and user charges.
Current position: Associate Professor at the Department of Economics, College of Economics and Management, University of the Philippines Los Baños (UPLB), Laguna, Philippines.
Web page: http://www.uplb.edu.ph
E-mail: maocatelo@uplb.edu.ph, maocatelo@gmail.com

Roberta Centonze

Expertise: Agricultural Economics and Policy, Sustainable Agriculture and Rural Development. Her research deals with rural system appraisal, agrofood, chain management and intellectual property rights on genetic resources.
Current position: Consultant in Agricultural Economics, Project Design and Management, University-Enterprise cooperation programmes, Bologna, Italy.
E-mail: roberta.centonze@fastwebnet.it

Chen Tong

Expertise: Agricultural Economics. Current research interests regard rural development economics, farm economics and management, agribusiness economics and ecological management. His research focuses on topics related to the Xinjiang province of China.
Current position: Professor of Agricultural Economics at the Xinjiang Academy of Agricultural Science. He is Vice President of the Chinese Association of Agricultural Economists, President of the Xinjiang Association of Agricultural Economists, President of the Xinjiang Association of Agriculture, Vice-Chairman of the Xinjiang Association of Science and Technology, Vice-President of the Xinjiang Statistics Association, Member of the Standing Committee of the China Society of Agrotechnical Economics, Urumqi, Xinjiang, China.
E-mail: ctelay2005@163.com, ctelay@163.com

Achilles Costales

Expertise: Agricultural economics, livestock economics; research and applications in design of public policy and institutions to efficiently and effectively integrate smallholder livestock keepers and stakeholders along supply chains in domestic markets as a means toward poverty alleviation among rural households in developing countries.
Current position: Livestock Economist of the Pro-Poor Livestock Policy Initiative (PPLPI) at the Food and Agriculture Organization (FAO) of the UN, Agriculture Department, Animal Production and Health Division (AGA), Rome, Italy.
Web page: http://www.fao.org/ag/pplpi.html
E-mail: achilles.costales@fao.org, achilles.costales@gmail.com

Ika Darnhofer

Expertise: Farm management and decision processes on family farms; conversion to organic farming; resilience, adaptability and transformability of family farms.
Current position: Associate Professor at the Institute of Agricultural and Forestry Economics, Department of Economics and Social Sciences, University of Natural Resources and Applied Life Sciences, Vienna (BOKU), Austria.
Web page: http://www.wiso.boku.ac.at/darnhofer.html
E-mail: ika.darnhofer@boku.ac.at

Wyn Ellis

Expertise: Organic agriculture policy, accreditation, certification and production; organic agriculture as a component of 'Green Growth' movements; agro-innovation systems.
Current position: Consultant and Senior Programme Adviser, Thai-German Programme for Enterpreise Competitiveness, GTZ Thailand. Vice-President, Organic Agriculture Foundation of Thailand, Bangkok, Thailand.
E-mail: wynellis.gtzbkk@gmail.com

Diana Feliciano

Expertise: Forest owners' organisations, forest policies, rural development. PhD topic: The contribution of the rural use sector to greenhouse gas neutral regions.
Current position: PhD student at ACES - The University of Aberdeen & The Macaulay Institute, The School of Biological Sciences, Aberdeen, UK
Web page: http://www.aces.ac.uk/people/#Diana_Feliciano, http://www.macaulay.ac.uk/staff/staffdetails.php?dianafeliciano
E-mail: d.feliciano@macaulay.ac.uk

Rainer Haas

Expertise: Consumer behaviour concerning organic and functional food. E-business and e-collaboration in the food value chain. Application of decision support systems (Analytic Hierarchy Process) in the field of food marketing.
Current position: Associate Professor at the Institute of Marketing & Innovation, Department of Economics and Social Sciences, University of Natural Resources and Applied Life Sciences, Vienna (BOKU), Austria.
Web page: http://www.wiso.boku.ac.at/mi_haas.html
E-mail: rainer.haas@boku.ac.at

About the authors

Huliyeti Hasimu

Expertise: Agri-food Economics, trade and marketing. Recent research subjects are focused on quality food products such as organic food and geographical indications for agricultural products.
Current position: Associate Professor at the Department of Economics and Trade, Xinjiang Agricultural University, Urumqi, Xinjiang (China) and project manager at Bioagrico-op scarl, Casalecchio di Reno (Italy).
E-mail: oriet@sina.com

Alexander Kasterine

Expertise: Agricultural and environmental economics, trade and climate change; organic agriculture, biodiversity and market incentives; carbon markets; product carbon footprint labelling.
Current position: Senior Adviser (Trade and Environment), International Trade Centre (UNCTAD/WTO), Geneva, Switzerland.
Web page: http://www.intracen.org/organics
E-mail: kasterine@intracen.org

Liu Ruifeng

Expertise: Agricultural Economics. Current research interests regard the organic food and GIs agricultural industry. His research focused on topics related to the Xinjiang province of China. His doctoral dissertation topic is 'Study on Production, Consumption and Policy Effect of Geographical Indications Agricultural Product in Xinjiang'.
Current position: PhD student at the College of Economics and Management, Xinjiang Agricultural University, Xinjiang (China).
E-mail: ruifeng076@sina.com

Pamela Lombardi

Expertise: Research focus on marketing researches in agri-food sector (qualitative and quantitative techniques) and on certified quality management systems in agri-food sector.
Current position: PhD student at the Department of Agricultural Economics and Engineering, Faculty of Agriculture, Alma Mater Studiorum - University of Bologna (UNIBO), Bologna, Italy.
E-mail: pam.lombardi@gmail.com

Mai Thach Hoanh

Expertise: Food crops, in-situ conservation of plant genetic resources, expert for taro and sweet potato.
Current position: Associate Professor, Researcher and consultant, Plant Resources Center of Vietnam, An Khanh, Dan Phuong district, Hanoi - Vietnam Academy of Agricultural Sciences (VAAS), Thanh Tri district, Hanoi, Vietnam.
Web page: http://www.pgrvietnam.org.vn
E-mail: maihoanh2006@yahoo.com

Sergio Marchesini

Expertise: Research focus on agricultural marketing in Asia.
Current position: PhD student at the Department of Agricultural Economics and Engineering, Faculty of Agriculture, Alma Mater Studiorum - University of Bologna (UNIBO), Bologna, Italy.
E-mail: smarch78@gmail.com

Floriana Marin

Expertise: Environmental economics, qualitative analysis. Research focus on public perception of genetically modified food and technology-related risks, as well as social studies of science.
Current position: Communication manager at the Research and Innovation Centre, Fondazione Edmund Mach Iasma, S. Michele all'Adige (TN), Italy.
Web page: http://research.iasma.it
E-mail: floriana.marin@iasma.it

Ellen McCullough

Expertise: The role of agriculture in poverty reduction, The role of policy in enabling agricultural productivity growth, Transformation of agri-food systems, Agriculture and nutrition linkages.
Current position: Associate Program Officer in the Agricultural Development initiative, Bill & Melinda Gates Foundation, Seattle, USA.
Web page: http://www.gatesfoundation.org
E-mail: ellen.mccullough@gatesfoundation.org

Ulrich B. Morawetz

Expertise: Research focus on econometric modelling of agricultural production.
Current position: PhD Student at the Institute of Sustainable Economic Development, Department of Economics and Social Sciences, University of Natural Resources and Applied Life Sciences, Vienna (BOKU), Austria.
Web page: http://homepage.boku.ac.at/umorawet/
E-mail: ulrich.morawetz@boku.ac.at

Nguyen Thanh Lam

Expertise: Agroecology, Upland farming, *in-situ* conservation of plant genetic resources, climate change and its impact on agricultural production.
Current position: Executive Director at the Center for Agricultural Research and Ecological Studies (CARES), Hanoi University of Agriculture, Hanoi, Vietnam.
Web page: http://www.cares.org.vn
E-mail: lamnt@cares.org.vn

Gianluca Nigro

Expertise: Its scientific interests are mainly the analysis of the costs of production and the economic sustainability of organic crops and no food (bioethanol, biodiesel, biomass, biogas, cosmetics, textiles).
Current position: Consultant at the C.I.C.A. Bologna (Interprovincial Consortium of Agricultural Cooperatives), in charge of economic studies and company management systems for the agri-food industry, Bologna, Italy.
Web page: http://www.cicabo.it
E-mail: g.nigro@cicabo.it

Stefano Padulosi

Expertise: Agriculturalist and biologist; research focus on plant genetic resources, conservation methods and sustainable use; specialized in the area of neglected and underutilized species and their deployment for enhancing people's livelihood.
Current position: Senior scientist at Bioversity International, Agency of the CGIAR headquartered in Rome (Italy); coordinating the organization's efforts dealing with agrobiodiversity and human wellbeing; leading a global multi-disciplinary UN Programme on neglected and underutilized species, focusing in Latin America, South and West Asia.
Web page: http://www.bioversityinternational.org
E-mail: s.padulosi@cgiar.org

Vitoon R. Panyakul

Expertise: Organic agricultural production and marketing. Supply chain management of organic products, especially rice, vegetables and fruits. Guarantee system of organic agriculture, from internal control system, to certification and accreditation.
Current position: General Secretary of the Earth Net Foundation / Green Net, Bangkok, Thailand.
Web page: http://www.greennet.or.th
E-mail: vitoon@greennet.or.th

Maria Papadopoulou

Expertise: Research focus on the motivations for the adoption of GlobalGAP in Emilia, Romagna in the fruit and vegetable sector; fruits and vegetables quality control; sensory analysis.
Current position: PhD student at the Department of Agricultural Economics and Engineering, Faculty of Agriculture, Alma Mater Studiorum - University of Bologna (UNIBO), Bologna, Italy. Quality control employee at FQC Italia, Casalecchio di Reno, Italy.
E-mail: maria.papadopoulou@gmail.com

Erika Pignatti

Expertise: Research focus on quality management systems and certification systems in the food industry, organic food certification systems and marketing.
Current position: Junior Research Fellow at the Department of Agricultural Economics and Engineering, Faculty of Agriculture, Alma Mater Studiorum - University of Bologna (UNIBO), Bologna, Italy.
Web page: http://www.deiagra.unibo.it/
E-mail: erika.pignatti@unibo.it

Prabhu Pingali

Expertise: Technical change in developing country agriculture, Agricultural development and food security, Biotechnology, Impact assessment and priority setting, Environmental sustainability, Post-Green Revolution Asian agriculture, Agricultural intensification in sub-Saharan Africa, Transformation of agri-food systems.
Current position: Deputy Director of Agricultural Development, Bill & Melinda Gates Foundation, Seattle, USA.
Web page: http://www.gatesfoundation.org
E-mail: prabhu.pingali@gatesfoundation.org

Siegfried Pöchtrager

Expertise: Marketing and market research for the food value chain. Total Quality Management, International Food Standard, Organic production standards, Quality management system – Requirements ISO 9001 – 2008, Food Safety management systems, Requirements for organizations throughout the food chain ISO 22000:2004, Traceability - Reg. (EC) 178/2002, Hazard Analysis Critical Control Point-Concept.
Current position: Researcher, Lecturer at the Institute of Marketing & Innovation, Department of Economics and Social Sciences, University of Natural Resources and Applied Life Sciences, Vienna (BOKU), Austria.
Web page: www.boku.ac.at/mi
E-mail: poechtrager@boku.ac.at

Bettina Riedel

Expertise: Food marketing, food value chain and food cluster studies. Research on competitiveness of European fresh vegetable business, with special focus on regional production organization.
Current position: PhD student at the Institute of Horticultural Economics, Department of Agricultural Economics, Humboldt-University of Berlin, Berlin, Germany.
Web page: http://www.agrar.hu-berlin.de/struktur/institute/wisola/fg/gp/standardseite
E-mail: riedel.bettina@gmail.com

Birgit Roitner-Schobesberger

Expertise: Organic food and farming, marketing of farm products; Education and training on organic food and regional products; Promoting underutilized crops such as grain amaranth.
Current position: PhD student at the Institute of Agronomy and Plant Breeding, University of Natural Resources and Applied Life Sciences, Vienna (BOKU) and Project coordinator at the Agrar.Projekt.Verein, Austria.
E-mail: roitner-schobesberger@boku.ac.at

Markus Schmidt

Expertise: Research focus on management of plant genetic resources for food and agriculture in Europe and Asia, promotion of neglected and underutilized crop species, technology assessment of novel biotechnologies such as synthetic biology.
Current position: Leader of the Biosafety Working group at the Organisation for International Dialogue and Conflict Management, based in Vienna, Austria. Project websites: http://www.diverseeds.eu (DIVERSEEDS: management of plant genetic resources in Europe and Asia). http://www.agrofolio.eu (AGRO-FOLIO: Benefiting from an improved agricultural portfolio in Asia)
Web page: http://www.idialog.eu
E-mail: markus.schmidt@idialog.eu

Bill Slee

Expertise: Rural economist specialising in the economics of land use, rural socio-economic change and sustainable development of rural areas. He has worked extensively in farm adjustment, agri-environmental issues, forestry and rural development, environment and development and rural tourism for a range of national and international clients including the World Bank, the European Commission, DFID and OECD, as well as a range of EU agencies.

Current position: Science Group Leader of the Socio-Economics Research Group at the Macaulay Institute. He was previously Professor of Rural Economy and Director of the Countryside and Community Research Unit in Cheltenham and, prior to that, a Senior Lecturer in the University of Aberdeen, UK.

Web page: http://www.macaulay.ac.uk/staff/staffdetails_billslee.html

E-mail: b.slee@macaulay.ac.uk

Suthichai Somsook

Expertise: Entomopathogenic Nematode, Organic Farming.

Current position: Associate Professor at the Department of Agricultural Technology, Faculty of Science and Technology, Thammasat University, Rangsit Campus, Pathumthani, Thailand.

Web page: http://www.tu.ac.th/eng/faculty/science_technology/

E-mail: suthisomsook@hotmail.com

Roberta Spadoni

Expertise: Economics of agricultural and food markets. Research focus on certification systems, agricultural and industrial marketing, product quality issues.

Current position: Assistant professor (tenured) at the Department of Agricultural Economics and Engineering, Faculty of Agriculture, Alma Mater Studiorum - University of Bologna (UNIBO), Bologna, Italy.

Web page: http://www.unibo.it/docenti/roberta.spadoni

E-mail: roberta.spadoni@unibo.it

Kostas Stamoulis

Expertise: The role of agriculture in growth and poverty reduction, Effects of stabilization on agricultural performance, The role of the non-farm rural sector on growth and food security, Impacts of urbanization and globalization.

Current position: Director, Agricultural Development Economics Division, Food and Agriculture Organization of the United Nations, Rome, Italy.

Web page: http://www.fao.org/es/esa

E-mail: Kostas.Stamoulis@fao.org

Christian R. Vogl

Expertise: Organic farming and traditional land use systems; Standards, regulations, inspection, auditing and accreditation on organic farming; organic farming in tropical and subtropical areas, local and indigenous knowledge of organic farmers, organic farmers' experiments, ethnobiology, biocultural diversity.

Current position: Associate Professor at the Division of Organic Farming, Department Sustainable Agriculture Systems, University of Natural Resources and Applied Life Sciences, Vienna (BOKU), Austria.

Web page: http://www.nas.boku.ac.at/christian-vogl.html

E-mail: christian.vogl@boku.ac.at

Rungsaran Wongprawmas

Expertise: Natural Resources Management and Agri-food marketing. Research focus on organic farming management regard to farm economic and financial analysis and Geographical Indications products with regard to gate-keepers and international markets issues.

Current position: Graduate student, International Master in Horticultural Sciences, Faculty of Agriculture, Alma Mater Studiorum - University of Bologna (UNIBO), Bologna, Italy.

E-mail: rungsaran.wongprawmas80@gmail.com

Keyword index

Printed in the United States
by Baker & Taylor Publisher Services